Lecture Notes in Computer Science 12406

More information about this series at http://www.springer.com/series/7408

Wei-Shinn Ku · Yasuhiko Kanemasa ·
Mohamed Adel Serhani ·
Liang-Jie Zhang (Eds.)

Web Services – ICWS 2020

27th International Conference
Held as Part of the Services Conference Federation, SCF 2020
Honolulu, HI, USA, September 18–20, 2020
Proceedings

Springer

Editors
Wei-Shinn Ku 🔘
Auburn University
Auburn, AL, USA

Yasuhiko Kanemasa
Fujitsu (Japan)
Kanagawa, Japan

Mohamed Adel Serhani 🔘
United Arab Emirates University
Al Ain, United Arab Emirates

Liang-Jie Zhang 🔘
Kingdee International Software
Group Co., Ltd.
Shenzhen, China

ISSN 0302-9743   ISSN 1611-3349   (electronic)
Lecture Notes in Computer Science
ISBN 978-3-030-59617-0   ISBN 978-3-030-59618-7   (eBook)
https://doi.org/10.1007/978-3-030-59618-7

LNCS Sublibrary: SL2 – Programming and Software Engineering

This Springer imprint is published by the registered company Springer Nature Switzerland AG
The registered company address is: Gewerbestrasse 11, 6330 Cham, Switzerland

Preface

The International Conference on Web Services (ICWS) has been a prime international forum for both researchers and industry practitioners to exchange the latest fundamental advances in the state of the art and practice of Web-based services, identifying emerging research topics and defining the future of Web-based services. All topics regarding Internet/Web services lifecycle study and management align with the theme of ICWS.

ICWS 2020 is a member of the Services Conference Federation (SCF). SCF 2020 had the following 10 collocated service-oriented sister conferences: the International Conference on Web Services (ICWS 2020), the International Conference on Cloud Computing (CLOUD 2020), the International Conference on Services Computing (SCC 2020), the International Conference on Big Data (BigData 2020), the International Conference on AI & Mobile Services (AIMS 2020), the World Congress on Services (SERVICES 2020), the International Conference on Internet of Things (ICIOT 2020), the International Conference on Cognitive Computing (ICCC 2020), the International Conference on Edge Computing (EDGE 2020), and the International Conference on Blockchain (ICBC 2020). As the founding member of SCF, the First International Conference on Web Services (ICWS 2003) was held in June 2003 in Las Vegas, USA. Meanwhile, the First International Conference on Web Services - Europe 2003 (ICWS-Europe 2003) was held in Germany in October 2003. ICWS-Europe 2003 was an extended event of ICWS 2003, and held in Europe. In 2004, ICWS-Europe was changed to the European Conference on Web Services (ECOWS), which was held in Erfurt, Germany.

This volume presents the accepted papers of ICWS 2020, held virtually during September 18–20, 2020. For this conference, each paper was reviewed by three independent members of the International Program Committee. After carefully evaluating their originality and quality, we accepted 14 papers.

We are pleased to thank the authors whose submissions and participation made this conference possible. We also want to express our thanks to the Organizing Committee and Program Committee members for their dedication in helping to organize the conference and reviewing the submissions. We owe special thanks to the keynote speakers for their impressive speeches.

Finally, we would like to thank operations team members, Dr. Sheng He and Dr. Yishuang Ning, for their excellent work in organizing this conference. We thank all volunteers, authors, and conference participants for their great contributions to the fast-growing worldwide services innovations community.

July 2020

Wei-Shinn Ku
Yasuhiko Kanemasa
Mohamed Adel Serhani
Noseong Park
Liang-Jie Zhang

Organization

General Chairs

Cheng-Zhong Xu   University of Macau, Macau
Raj Sunderraman   Georgia State University, USA

Program Chairs

Wei-Shinn Ku   National Science Foundation, USA
Yasuhiko Kanemasa   Fujitsu Ltd, Japan
Mohamed Adel Serhani   United Arab Emirates University, UAE
Noseong Park (Vice-chair)   George Mason University, USA

Services Conference Federation (SCF 2020)

General Chairs

Yi Pan   Georgia State University, USA
Samee U. Khan   North Dakota State University, USA
Wu Chou   Vice President of Artificial Intelligence & Software at Essenlix Corporation, USA
Ali Arsanjani   Amazon Web Services (AWS), USA

Program Chair

Liang-Jie Zhang   Kingdee International Software Group Co., Ltd, China

Industry Track Chair

Siva Kantamneni   Principal/Partner at Deloitte Consulting, USA

CFO

Min Luo   Georgia Tech, USA

Industry Exhibit and International Affairs Chair

Zhixiong Chen   Mercy College, USA

Operations Committee

Jing Zeng   Yundee Intelligence Co., Ltd, China
Yishuang Ning   Tsinghua University, China
Sheng He   Tsinghua University, China
Yang Liu   Tsinghua University, China

Steering Committee

Calton Pu (Co-chair)	Georgia Tech, USA
Liang-Jie Zhang(Co-chair)	Kingdee International Software Group Co., Ltd, China

ICWS 2020 Program Committee

Adel Abusitta	McGill University, Canada
Ismailcem Arpinar	University of Georgia, USA
Yacine Atif	University of Skövde, Sweden
Elarbi Badidi	United Arab Emirates University, UAE
Amin Beheshti	Macquarie University, Australia
Ladjel Bellatreche	LISI/ENSMA University of Poitiers, France
Fatna Belqasmi	Zayed University, UAE
Nizar Bouguila	Concordia University, Canada
May El Barachi	University of Wollongong in Dubai, UAE
Abdelghani Benharref	University of Wollongong in Dubai, UAE
Hadeel El-Kassabi	United Arab Emirates University, UAE
Miki Enoki	IBM Research Tokyo, Japan
Marios-Eleftherios Fokaefs	Polytechnique Montréal, Canada
Walid Gaaloul	Télécom SudParis, France
Keke Gai	Beijing Institute of Technology, China
Hakim Hacid	Zayed University, UAE
Saad Harous	United Arab Emirates University, UAE
Abdul Kadhim Hayawi	Zayed University, UAE
Farkhund Iqbal	Zayed University, UAE
Gueyoung Jung	AT&T Labs, USA
Hyuk-Yoon Kwon	Seoul National University of Science & Technology, South Korea
Abderrahmane Lakas	United Arab Emirates University, UAE
Jaehwan Lee	Korea Aerospace University, South Korea
Jing Liu	George Mason University, USA
Zakaria Maamar	Zayed University, UAE
Sujith Mathew	Zayed University, UAE
Massimo Mecella	Sapienza Università di Roma, Italy
Rabeb Mizouni	Khalifa University, UAE
Mahmoud Mohammadi	Lowe's Companies Inc., USA
Hadi Otrok	Khalifa University, UAE
Ali Ouni	ÉTS Montreal, Canada
Young-Kyoon Suh	Kyungpook National University, South Korea
Ikbal Taleb	Zayed University, UAE
Liqiang Wang	University of Central Florida, USA
Lei Yang	South China University of Technology, China
Lina Yao	The University of New South Wales, Australia
Yao Yu	The University of Sydney, Australia
Rui Zhang	Institute of Information Engineering, Chinese Academy of Sciences, China

Yang Zhou   Auburn University, USA
Salah Bouktif   United Arab Emirates University, UAE
Heba Ismail   United Arab Emirates University, UAE
Alramzana N. Navaz   United Arab Emirates University, UAE

Conference Sponsor – Services Society

Services Society (S2) is a nonprofit professional organization that has been created to promote worldwide research and technical collaboration in services innovation among academia and industrial professionals. Its members are volunteers from industry and academia with common interests. S2 is registered in the USA as a "501(c) organization," which means that it is an American tax-exempt nonprofit organization. S2 collaborates with other professional organizations to sponsor or co-sponsor conferences and to promote an effective services curriculum in colleges and universities. The S2 initiates and promotes a "Services University" program worldwide to bridge the gap between industrial needs and university instruction.

The services sector accounted for 79.5% of the USA's GDP in 2016. The world's most service-oriented economy, with services sectors accounting for more than 90% of GDP. S2 has formed 10 Special Interest Groups (SIGs) to support technology and domain specific professional activities.

- Special Interest Group on Web Services (SIG-WS)
- Special Interest Group on Services Computing (SIG-SC)
- Special Interest Group on Services Industry (SIG-SI)
- Special Interest Group on Big Data (SIG-BD)
- Special Interest Group on Cloud Computing (SIG-CLOUD)
- Special Interest Group on Artificial Intelligence (SIG-AI)
- Special Interest Group on Edge Computing (SIG-EC)
- Special Interest Group on Cognitive Computing (SIG-CC)
- Special Interest Group on Blockchain (SIG-BC)
- Special Interest Group on Internet of Things (SIG-IOT)

About the Services Conference Federation (SCF)

As the founding member of the Services Conference Federation (SCF), the First International Conference on Web Services (ICWS 2003) was held in June 2003 in Las Vegas, USA. Meanwhile, the First International Conference on Web Services - Europe 2003 (ICWS-Europe 2003) was held in Germany in October 2003. ICWS-Europe 2003 was an extended event of ICWS 2003, and held in Europe. In 2004, ICWS-Europe was changed to the European Conference on Web Services (ECOWS), which was held in Erfurt, Germany. SCF 2019 was held successfully in San Diego, USA. To celebrate its 18th birthday, SCF 2020 was held virtually during September 18–20, 2020.

In the past 17 years, the ICWS community has been expanded from Web engineering innovations to scientific research for the whole services industry. The service delivery platforms have been expanded to mobile platforms, Internet of Things (IoT), cloud computing, and edge computing. The services ecosystem is gradually enabled, value added, and intelligence embedded through enabling technologies such as big data, artificial intelligence (AI), and cognitive computing. In the coming years, all the transactions with multiple parties involved will be transformed to blockchain.

Based on the technology trends and best practices in the field, SCF will continue serving as the conference umbrella's code name for all service-related conferences. SCF 2020 defines the future of New ABCDE (AI, Blockchain, Cloud, big Data, Everything is connected), which enable IOT and enter the 5G for Services Era. SCF 2020's 10 collocated theme topic conferences all center around "services," while each focusing on exploring different themes (web-based services, cloud-based services, big data-based services, services innovation lifecycle, AI-driven ubiquitous services, blockchain driven trust service-ecosystems, industry-specific services and applications, and emerging service-oriented technologies). SCF includes 10 service-oriented conferences: ICWS, CLOUD, SCC, BigData Congress, AIMS, SERVICES, ICIOT, EDGE, ICCC, and ICBC. The SCF 2020 members are listed as follows:

[1] The International Conference on Web Services (ICWS 2020, http://icws.org/) is the flagship theme-topic conference for Web-based services, featuring Web services modeling, development, publishing, discovery, composition, testing, adaptation, delivery, as well as the latest API standards.

[2] The International Conference on Cloud Computing (CLOUD 2020, http://thecloudcomputing.org/) is the flagship theme-topic conference for modeling, developing, publishing, monitoring, managing, delivering XaaS (Everything as a Service) in the context of various types of cloud environments.

[3] The International Conference on Big Data (BigData 2020, http://thecloudcomputing.org/) is the emerging theme-topic conference for the scientific and engineering innovations of big data.

[4] The International Conference on Services Computing (SCC 2020, http://thescc.org/) is the flagship theme-topic conference for services innovation lifecycle that includes enterprise modeling, business consulting, solution creation, services

orchestration, services optimization, services management, services marketing, and business process integration and management.

[5] The International Conference on AI & Mobile Services (AIMS 2020, http://thescc.org/) is the emerging theme-topic conference for the science and technology of AI, and the development, publication, discovery, orchestration, invocation, testing, delivery, and certification of AI-enabled services and mobile applications.

[6] The World Congress on Services (SERVICES 2020, http://servicescongress.org/) focuses on emerging service-oriented technologies and the industry-specific services and solutions.

[7] The International Conference on Cognitive Computing (ICCC 2020, http://thecognitivecomputing.org/) focuses on the Sensing Intelligence (SI) as a Service (SIaaS) which makes systems listen, speak, see, smell, taste, understand, interact, and walk in the context of scientific research and engineering solutions.

[8] The International Conference on Internet of Things (ICIOT 2020, http://iciot.org/) focuses on the creation of IoT technologies and development of IoT services.

[9] The International Conference on Edge Computing (EDGE 2020, http://theedgecomputing.org/) focuses on the state of the art and practice of edge computing including but not limited to localized resource sharing, connections with the cloud, and 5G devices and applications.

[10] The International Conference on Blockchain (ICBC 2020, http://blockchain1000.org/) concentrates on blockchain-based services and enabling technologies.

Some highlights of SCF 2020 are shown below:

- **Bigger Platform:** The 10 collocated conferences (SCF 2020) are sponsored by the Services Society which is the world-leading nonprofit organization (501 c(3)) dedicated to serving more than 30,000 worldwide services computing researchers and practitioners. Bigger platform means bigger opportunities to all volunteers, authors, and participants. Meanwhile, Springer sponsors the Best Paper Awards and other professional activities. All the 10 conference proceedings of SCF 2020 have been published by Springer and indexed in ISI Conference Proceedings Citation Index (included in Web of Science), Engineering Index EI (Compendex and Inspec databases), DBLP, Google Scholar, IO-Port, MathSciNet, Scopus, and ZBlMath.
- **Brighter Future:** While celebrating the 2020 version of ICWS, SCF 2020 highlights the Third International Conference on Blockchain (ICBC 2020) to build the fundamental infrastructure for enabling secure and trusted service ecosystems. It will also lead our community members to create their own brighter future.
- **Better Model:** SCF 2020 continues to leverage the invented Conference Blockchain Model (CBM) to innovate the organizing practices for all the 10 theme conferences.

Contents

A Reputation Based Hybrid Consensus for E-Commerce Blockchain

You Sun[1,2], Rui Zhang[1,2(✉)], Rui Xue[1,2], Qianqian Su[1,2], and Pengchao Li[1]

[1] State Key Laboratory of Information Security,
Institute of Information Engineering,
Chinese Academy of Sciences, Beijing 100093, China
{sunyou,zhangrui,xuerui,suqianqian,lipengchao}@iie.ac.cn
[2] School of Cyber Security, University of Chinese Academy of Sciences,
Beijing 100049, China

Abstract. Blockchain can achieve non-tampering, non-repudiation, consistency and integrity that other data management technologies do not have. Especially in peer-to-peer networks, the decentralized nature of blockchain has drawn tremendous attention from academic and industrial communities. Recently, the field of e-commerce has also begun to realize its important role. Although blockchain technology has many advantages in achieving trust establishment and data sharing among distributed nodes, in order to make it better to be applied in e-commerce, it is necessary to improve the security of transactions and the efficiency of consensus mechanisms. In this paper, we present a reputation based hybrid consensus to solve the problem of transaction security and efficiency. Our scheme integrates the reputation mechanism into transactions and consensus, and any improper behavior of nodes will be reflected in the reputation system and fed back to a new round of transactions and consensus. We implement distributed reputation management and enable users to append new reputation evaluations to the transaction that has previously evaluated. Meanwhile, we demonstrated that the scheme can defend against existing attacks such as selfish mining attacks, double spending attacks and flash attacks. We implement a prototype and the result shows that our scheme is promising.

Keywords: Blockchain · E-commerce · Consensus mechanism · Reputation system

1 Introduction

Recent advances in blockchain technology have witnessed unprecedented practicability in various tasks e.g., intelligent services, health care, education, social management. The consensus mechanism is the core technology of the blockchain, and mainstream blockchain platforms such as Bitcoin and Ethereum rely on the continuous mining of miners distributed around the world to maintain the normal operation of the system. However, the mining incentive mechanism will lead

© Springer Nature Switzerland AG 2020
W.-S. Ku et al. (Eds.): ICWS 2020, LNCS 12406, pp. 1–16, 2020.
https://doi.org/10.1007/978-3-030-59618-7_1

to the concentration of computing power, which will not only cause 51% attacks but also waste a lot of resources. The security of consensus mechanisms such as proof-of-stake (POS) [7,14,24] and delegated proof-of-stake (DPOS) [1], which is not based on computing power, has not been theoretically validly proven. Strong consistency algorithms such as practical Byzantine fault-tolerant algorithms also have disadvantages such as high algorithm complexity and low degree of decentralization. Therefore, to apply blockchain technology to the field of e-commerce, how to design a safe and efficient consensus mechanism is one of the major challenges.

In this paper, we propose a hybrid consensus mechanism and apply it to the e-commerce blockchain. We have put forward the concept of reputation. The purpose of this is that, on the one hand, both parties to the transaction need the reputation system as an important basis for mutual judgment; on the other hand, integrating the reputation mechanism into the consensus can promote the normal operation of the consensus mechanism and made the system more secure.

To summarize, we made the following contributions:

- We introduced the blockchain into the e-commerce system to realize the decentralized management of transactions, so that users can complete transactions without the involvement of trusted third parties. We propose a reputation based hybrid consensus for e-commerce blockchain. We introduce the concept of reputation, and push forward nodes to act legally in both transactions and consensus.
- In our consensus mechanism, we decouple transaction packaging and write transaction lists into the blockchain. Besides, the work of miners is divided into low difficulty microblocks and high difficulty blocks, which effectively prevents the concentration of computing power and ensures the degree of decentralization of the system.
- We use a reputation chain to store, manage, and update reputations. In this way, distributed reputation management is achieved without the need for trusted third parties in traditional reputation systems. In addition, the reputation system also supports users to append new evaluation messages to the transaction that has already evaluated.
- We provide detailed security analysis of our scheme, the analysis shows that our scheme can resist the most attacks. We implement the prototype and evaluate its performance. The experiment result shows that our scheme has high efficiency, that is, it can achieve high throughput, so the scheme is more suitable to the field of e-commerce.

2   Related Work

With the wide application of blockchain technology, the types of consensus mechanisms have also become diverse to accommodate a variety of different application scenarios. In this section, we will give an introduction to the consensus mechanisms in different blockchains.

Proof of Work (POW) is the earliest consensus mechanism in the blockchain field, it serves the bitcoin network proposed by Nakamoto [10,19]. The work here refers to the process of computer computing calculating a random number. Within a certain period of time, the difficulty of finding a random number is certain, which means that it takes a certain amount of work to get this random number. The node that first obtains this random number is responsible for packing the transactions to the block, adding the new block to the existing blockchain, and broadcasting it to the entire network, other nodes perform verification and synchronization. In the case of POS, the system allocates the corresponding accounting rights according to the product of the number of tokens held by the node and the time. In DPOS, the person who owns the token votes to a fixed node, and these nodes act as agents to exercise the right to record. These voting-recognized representatives obtain billing rights in turn according to certain algorithms. PBFT is a fault-tolerant algorithm for Byzantine generals [16,18], it is a state machine that requires all nodes to jointly maintain a state, and all nodes take the same action. The Paxo mechanism without considering Byzantine faults [15] and Raft mechanism [20] belong to the same type of consensus mechanism. There are also some problems with these common consensus mechanisms, such as the waste of the computing power of the PoW mechanism, the interest of the PoS mechanism is easy to concentrate on the top layer, and the efficiency of the PBFT in the network with a large number of nodes that are constantly changing dynamically, etc.

Therefore, in recent years, scholars have been proposing new consensus mechanisms. Ouroboros [13] is the first blockchain protocol based on PoS with strong security guarantees, which has an efficiency advantage over a blockchain of physical resource proofs (such as PoW). Given this mechanism, honest behavior is an approximate Nash equilibrium, which can invalidate selfish mining attacks. Snow White [4] addresses the dynamic distribution of stake owners and uses corruption delay mechanisms to ensure security. Algorand [6] provides a distributed ledger that follows the Byzantine protocol, and each block resists adaptive corruption. Fruitchain [23] provides a reward mechanism and approximate Nash equilibrium proof for PoW-based blockchains.

Integrating identity and reputation into the consensus mechanism is a new direction. Proof of authority (PoA) is a reputation-based consistency algorithm that introduces practical and effective solutions for blockchain networks, especially private blockchains. The mechanism was proposed in 2017 by Ethereum co-founder Gavin Wood [2]. PoA uses the value of the identity, which means that the verifier of the block does not depend on the cryptocurrency of the mortgage but on the reputation of the individual. The PoA mechanism relies on a limited number of block validators, making it a highly scalable system. Blocks and transactions are verified by participants pre-approved by the system administrator. RepuCoin [25], developed by the University of Luxembourg's Interdisciplinary Centre for Security, Reliability and Trust, uses the concept of online reputation to defend against 51% of attacks. Gai et al. present a reputation-based consensus protocol called Proof of Reputation (PoR) [9], which guarantees the reliability and integrity of transaction outcomes in an efficient way.

However, the existing research results have not yet applied a sound reputation system on the e-commerce blockchain, and these consensus mechanisms or systems based on reputation do not consider the behavior of users in the transaction process, such as users do not comply with or deliberately destroy the transaction rules. Therefore, this paper combines consensus with actual transactions through a reputation mechanism, and proposes a reputation-based consensus mechanism suitable for the application scenarios of e-commerce.

3   Notations and Security Definitions

In this section, we will introduce some notations and the security definition in the scheme.

3.1   Notations

Towards to formalize our consensus mechanism and blockchain system, we first define some basic concepts and notations.

User. In our scheme, the public key pk_j is used to mark the user j. pk_j is public, but its corresponding sk_j is owned only by user j.

In the scheme, there are also some special users. The consensus group members reach a consensus on the transaction list. Each round has a leader who packs transactions in the network and initiates consensus. Miners generate corresponding blocks by solving puzzles of different difficulty. From the perspective of application scenarios of transactions, users are also divided into service providers and purchasers.

Transaction. In a transaction, the purchaser i first needs to send a service requirement to the service provider j. We define it as: $ServiceRequirement_{ij} = (pk_i, pk_j, Req, H(Req'), \sigma_i)$. where pk is the public key of the purchaser, pk' is the service provider's public key, Req is the specific service requirement, where the part involving sensitive information is represented by $H(Req')$, and σ is the purchaser's signature of the above information. After the service provider receives the requirement message, it provides the corresponding service to the purchaser, and then sends a response message to the purchaser: $ServiceResponse_{ij} = (ServiceRequirement_{ij}, Res, H(Res'), \sigma_j)$. After receiving the message and corresponding service, the purchaser adds the service result (the service result is 1 if the transaction is completed, otherwise 0) to the response message, signs it and publishes it to the blockchain network: $Service_{ij} = (ServiceResponse_{ij}, result, \sigma_i)$, this process can be expressed as Fig. 1.

There are some special transaction forms such as reputation transactions and reputation modification transactions, the specific structure of which will be introduced later.

A Transaction List is a collection of transactions of the same type. It is determined after the consensus group has reached a consensus, and it has an order. Therefore, in addition to the transactions in the network over a period of time, the transaction list also needs to include a serial number, the hash value of the previous transaction list, and the signatures of the consensus group members σ_T.

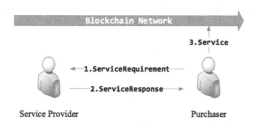

Fig. 1. The transaction generation process.

MicroBlock. Each MicroBlock corresponds to a transaction list, which is a block with lower mining difficulty. The existence of MicroBlock is to enable miners to receive certain rewards without having to aggregate a large amount of computing power, thereby preventing selfish mining. The main structure of Microblock is:

$$MicroBlock = (recent\_block\_hash, tran\_serial\_num, tranList\_hash, Nonce)$$

where $recent\_block\_hash$ is the hash value of the latest block on the current blockchain, $tran\_serial\_num$ is the serial number of the transaction list contained in the MicroBlock, and $tranList\_hash$ is its hash value. $Nonce$ is the solution to the puzzle.

Block. Block is similar to the structure in the Bitcoin system, and miners need to solve a puzzle to mine blocks. The difficulty of this puzzle is higher than that of MicroBlock. The main structure of a block is as follows:

$$Block = (prev\_block\_hash, MicroBlock, Nonce)$$

where $prev\_block\_hash$ is the hash value of the previous block, $MicroBlock$ is the collection of MicroBlocks contained in the Block, and $Nonce$ is the puzzle solution.

3.2   Security Definitions

In the scheme discussed in this paper, we follow the security assumptions of hybrid consensus mentioned in [21]. On the network side, the system is not a fully synchronized network, so in some cases messages on the network may be lost, duplicated, delayed, or out of order. We assume that messages are delivered within a certain time t. And the nodes in the system are independent, and the failure of one node will not cause the failure of other nodes.

We assume that there is a malicious adversary A in the system, and the computing power of adversary A is limited. For example, it cannot crack the encryption algorithm, forge the digital signature of a node, or recover the message content from the hash data. Adversary A cannot delay normal nodes indefinitely.

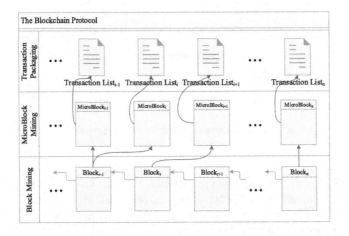

Fig. 2. The blockchain protocol.

Adversary A can manipulate multiple failed nodes, however, in the system, we assume that when there are f malicious nodes or failed nodes, then at least $2f + 1$ normal nodes must be guaranteed in the system. Only in this way can the security and liveness of the system in the consensus mechanism be guaranteed, and there will be no indefinite delay.

4   The Proposed Consensus Mechanism

In this section, we will introduce the main processes of the hybrid consensus mechanism. The consensus mechanism we propose is mainly divided into three stages, namely transaction packaging, microblock mining and block mining. Figure 2 describes the general flow of these three phases. We will describe it in detail below.

4.1   Transaction Packaging

The transaction packing phase loads the messages in the network into the transaction list in the recent period. Transaction packaging is done by members of the consensus group that are dynamically selected in each round. Members of the consensus group are some of the nodes with the highest reputation scores.

The calculation of consensus reputation will be described in detail later. Our scheme will select the leader with a $\frac{1}{|G|}$ probability among the members of the consensus group G with the highest reputation score, where $\frac{1}{|G|}$ is the number of consensus group members. The specific selection method of the r-th transaction list's leader l_r is as follows:

$$seed_r \longleftarrow H(TransactionList_{r-1})$$
$$i \longleftarrow seed_r \bmod |G|$$
$$l_r \longleftarrow G_i$$

The leader is responsible for packaging the messages and initiating consensus. It first sends the following messages to the members of the consensus group of this round: $(pk_{l_r}, TransactionList_r, SIG_{l_r}(TransactionList_r))$, after receiving the message, other consensus group members first check whether the sender is the leader of the current round, and if so, check whether the transaction list and its signature are legal. If all are legal, other nodes sign the message and broadcast to the consensus group. After a period of time, if members of the consensus group receive $2f$ messages that are the same as themselves, they send a commit message to the consensus group. After a while, if members of the consensus group receive $2f + 1$ commit messages, they will publish the transaction list. The form of the pinned transaction list is shown in Fig. 3.

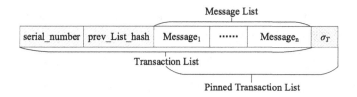

Fig. 3. The pinned transaction list structure.

The message list contains transaction messages published in the network over a period of time. When the leader packages it into a transaction list, it needs to add a serial number and the hash value of the previous pinned transaction list. After consensus is finally reached, members of the consensus group will attach their signature certificate to the transaction list to prove its validity.

4.2 MicroBlock Mining

In cryptocurrency systems such as Bitcoin, people have been worried about the concentration of computing power for a long time due to the existence of mining pools. Large mining pools and mining pool alliances composed of stakeholders may have close to or even exceed 51% of the computing power.

To solve this problem, we introduced MicroBlock. Each MicroBlock contains a transaction list. In addition, because MicroBlock can be mined out of order, it also needs to contain the serial number of the transaction list to restore the order of transactions. The mining difficulty of MicroBlock is relatively low, and it can be adjusted according to the operating status of the system. We hope that its mining interval is close to the consensus time of a transaction list. MicroBlock's puzzle is defined as follows:

$$H(prev\_block, recent\_block, Microblock\_set, tranList, Nonce) < target_m$$

where $prev\_block$ is the hash value of the previous Block on the current blockchain, although Microblock mining does not need to pay attention to it, in order to ensure that the miners solve the same puzzles as mining Blocks (but the difficulty is different), it needs to be added to the puzzle. $Microblock\_set$ represents the MicroBlocks contained in the Block. Like the former, it only plays a practical role in Block mining. $recent\_block$ is the Block on which this MicroBlock mining is based. It can be any Block within a period of time. The $tranList$ is the hash value of the transaction list included in the MicroBlock. It is worth noting that it should also contain the serial number of the transaction list. In the end, $Nonce$ is a solution to the puzzle, and the miner obtains the corresponding reward by finding $Nonce$.

When several MicroBlocks are mined at the same time, in this solution, the solution we choose is to choose the one with the smallest hash value.

4.3 Block Mining

Block mining is similar to the PoW-based mechanism in Bitcoin. In this scheme, the mining puzzle is defined as:

$$H(prev\_block, recent\_block, Microblock\_set, tranList, Nonce) < target$$

The mining difficulty $target$ is higher than MicroBlock's mining difficulty $target_m$, and it can also be adjusted during the system operation. Similar to MicroBlock, there are some things that Block does not care about in the puzzle of mining Blocks, such as $recent\_block$ and $tranList$. These are all related to mining MicroBlock. The former represents one of the most recent blocks that MicroBlock is attached to, and the latter records the transaction list stored in MicroBlock. $prev\_block$ is the hash of the previous Block, $Microblock\_set$ is the MicroBlocks included in the Block. A Block will contain multiple MicroBlocks. Block miners will pack as many MicroBlocks as possible to get more rewards.

4.4 Reputation System and Reward System

Reputation System. In this section, we mainly deal with the change of reputation in the consensus process. In addition, in the process of transactions, there will be changes in the transaction reputation, which we will discuss in detail in the next section.

Reputation feedback management can ensure the security of a decentralized system. Therefore, in our solution, reputation scores, as well as computing power, have become the key for miners to gain power in consensus mechanisms. We represent the reputation score of node i in the consensus as R_i. For the convenience of description, we show the notations used in the reputation system in Table 1, and reputation R is calculated according to Algorithm 1.

Table 1. The notations in reputation system

Notation	Explanation
α	A system parameter related to activity
b_i	The number of Blocks mined by node i
mb_i	The number of MicroBlocks mined by node i
L_i	The number of times node i becomes leader
$Init_i$	The initial reputation of node i
A_i	Activity of node i
t_i	The number of improper behaviors of node i
H_i	Honesty of node i

Algorithm 1. Reputation Calculation Algorithm

Input: the system parameter related to activity, α; the number of Blocks mined by node b_i; the number of MicroBlocks mined by node i, mb_i; the number of times node i becomes leader, L_i; the initial reputation, $Init_i$; the number of improper behaviors of node i , t_i;

Output: the consensus reputation score of node i, R_I;

1: Select the reputation system parameter α;

2: Calculate the number of times the node successfully mines and becomes a leader $p_i = b_i + mb_i + L_i$;

3: Calculate the activity $A_i = 1 - \dfrac{\alpha}{(p_i + Init_i) + \alpha}$;

4: Calculate the honesty $H_i = \dfrac{1}{1 + t_i}$;

5: $R_i = H_i \cdot A_i$;

    return R_i;

The consensus reputation of a node is determined by two factors: its activity and honesty. The activity of node i is mainly determined by three factors, which are the number of times a Block is mined by node i, the number of times a MicroBlock is mined by node i, and the number of times it becomes a leader. In addition, some nodes will have a high initial reputation, which depends on the specific scenarios of transactions in the real world. In the initial stage of the system, these nodes will play a key role in the consensus mechanism, but as the system runs, the node's reputation will be more affected by its behavior.

The honesty of a node depends on the legitimacy of its behavior. We define the following behaviors of nodes as improper behaviors:

- Nodes provide conflicting transactions when they act as leaders;
- Nodes as members of the consensus group submit messages that conflict with most members;
- Nodes do not include the correct information after mining Blocks or MicroBlocks.

When nodes have one of these behaviors, their honesty decreases. The selection of the consensus group in our consensus mechanism is to sort the node's reputation value R and select the highest $|G|$ nodes to participate in transaction packaging. Such a mechanism can encourage nodes to properly perform their work, thereby improving their reputation scores and obtaining more rewards.

Reward System. The reward mechanism is the core of the blockchain system and the largest source of power for the system's sustainable development. In our reward system, the main source is transaction fees and there are three types of nodes that receive rewards: leaders, block miners and microblock miners. For each final block, there is one block miner and multiple microblock miners and leaders. This is due to the structure of the block. A block contains multiple microblocks, and a microblock contains a transaction list.

Our reward system allocates transaction fees in the following way: For each transaction list, the leader of the transaction list, the miner of the microblock with this list, and the miner of the block where this microblock was eventually placed are divided equally the sum of transaction fees. That is, each node gets a third of the transaction fees. Such a reward system can prompt the leader to pack as many transactions as possible into the transaction list to get more rewards, and the same is true for block miners, who will be more inclined to put more microblocks into blocks. In addition, block miners will receive more rewards than microblocks, which is proportional to the computing power they pay.

The reward system and the reputation system have a mutually reinforcing role, because a node with the correct behavior can not only receive a reward for transaction fees, but also accumulate a higher reputation score, and increasing the reputation score can also increase the probability that the node will enter the consensus group and be selected as a leader.

5   The Reputation Chain

In this section, we will describe the management and storage of reputation in the transaction stage. We store the reputation scores in the transaction separately in the reputation chain, so that different nodes can choose the synchronized data according to their needs.

5.1   Reputation Storage and Management

In a transaction scenario, both parties to a transaction need a reputation system as an important basis for mutual evaluation. Therefore, we propose the concept of a reputation chain to perform distributed management of reputation evaluation in user transactions. In the reputation chain, there are two forms of reputation transactions, namely reputation publication transactions and additional reputation transactions.

After the node completes a transaction, it needs to evaluate the transaction. This type of message has the form:

$$repu\_pub = (pk_i, pk_j, tran\_location, tran\_hash, m_{pub}, \sigma)$$

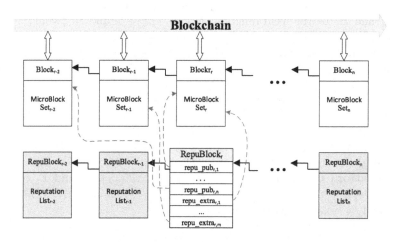

Fig. 4. The reputation chain structure.

where pk_i and pk_j are the public keys of the purchaser and the service provider respectively. $tran\_location$ is the position of the transaction corresponding to this reputation evaluation message in the transaction blockchain, and $tran\_hash$ is the hash of the transaction. m_{pub} is the content of reputation evaluation, σ is the publisher's signature on this message.

If the purchaser wants to publish additional reputation evaluation message after it has been submitted, the purchaser needs to generate a reputation modification message, which has the following structure:

$$repu\_extra = (pk_i, pk_j, tran\_location, tran\_hash, m_{extra}, \sigma')$$

The first four elements of the $repu\_extra$ message are the same as the $repu\_pub$, and they represent the same meaning. m_{extra} is the new content of the additional reputation evaluation message, and σ' is its signature on the message.

Figure 4 shows the structure of the reputation chain and its relationship with the original blockchain. Each of these blocks in the reputation chain is similar to $RepuBlock_r$, and contains two types of messages. The reputation evaluation message corresponds to a transaction in the original blockchain. The leader described in Sect. 4 is responsible for packaging transactions of the reputation chain and reaching a consensus within the consensus group.

5.2 Reputation and Additional Reputation Publication

Our scheme provides the ability to publish reputation evaluation message and additional reputation message to the same transaction and implements it using a linkable ring signature algorithm [12,17]. Before publishing reputation evaluation messages, nodes sign them in the following way:

- **[Setup]** Choose a cyclic additive group $\mathbb{G}$ which generator is p and a multiplicative group $\mathbb{G}_1$ of a large prime q. The bilinear pairing is $e : \mathbb{G} \times \mathbb{G} \to \mathbb{G}_1$. Let H and H' be the hash function, $H : \{0,1\}^* \to \mathbb{G}$ and $H' : \{0,1\}^* \to \mathbb{Z}_q$. Randomly choose $t \in mathbbZ_q$, $s \in mathbbZ_q^*$ and $A \in \mathbb{G}$, every legal user has an ID and let $S_{ID} = sH_i(ID)$ be the user's private key, its public key is given by $H'(ID)$, let $P_{pub} = sP$, $L = \{ID_i\}$. Compute $P' = tP$, $c_{k+1} = H(L \parallel m \parallel e(A,P) \parallel e(A,P))$.
- **[Generate ring sequence]** For $i = k+1, ..., n-1, 0, 1, ..., k-1$ (the value of i modulo n), choose $R_i, T_i \in \mathbb{G}$ randomly, and compute $c_{i+1} = H(L \parallel m \parallel e(R_i,P)e(H'(ID_i), P_{pub}) \parallel e(T_i, P)e(c_iH'(ID_i), P'))$.
- **[Forming the ring]** Let $R_k = A - c_kS_{ID_k}$ and $T_k = A - c_ktH'(ID_k)$.
- **[Signing]** the signature for m is $\sigma = (P', c_0, R_0, ..., R_{n-1}, T_0, ..., T_{n-1})$.
- **[Verification]** Given $(P', c_0, R_0, ..., R_{n-1}, T_0, ..., T_{n-1})$, m and L, compute $c_{i+1} = H(L \parallel m \parallel e(R_i,P), e(c_iH'(ID_i), P_{pub}) \parallel e(T_i, P)e(c_iH'(ID_i), P'))$. Accept if $c_n = c_0$, otherwise reject.

Table 2. Attack resilience

Attacks	BitCoin	ByzCoin	Hyperledger Fabric	Our scheme
Selfish mining attacks	×	×	√	√
Double spending attacks	×	√	√	√
Flash attacks	×	×	√	√
Reputation feedback	×	×	×	√

When the user needs to publish additional reputation evaluation information, he uses the same t as in the original message signature to front the new reputation evaluation message. Only from the same t, we can get the same P' as the original signature. When verifying the signature, a signature with the same P' is the additional message corresponding to the original message. If t' other than t is used for signing, the correctness of the signature algorithm cannot be guaranteed in the step of forming the ring. For non-signers, solving t is a discrete logarithm problem.

6 Security and Performance Analysis

In this section, we first analyze the security of the scheme, prove that our scheme can resist the existing attacks, and then we describe the efficiency of the scheme.

6.1 Security Analysis

Towards the current attacks, we give the security analysis for our consensus mechanism in this section and compared them with existing systems. The result is shown in Table 2.

Security Under Selfish Mining Attacks. Selfish mining [8] is mainly achieved by detaining blocks and delaying the time of publishing blocks. The purpose of selfish mining is not to destroy the blockchain network of cryptocurrencies, but to obtain greater profits. Because the threshold of attack is relatively low and the profit is high, theoretically this kind of attack is easy to appear. In the consensus mechanism described in this paper, miners are only responsible for putting as many microblocks containing transaction lists into the block as possible, and the block containing the most microblocks is the main chain. Therefore, miners can only publish the blocks they mine as soon as possible to get the maximum benefit, it cannot mine the block and not publish it, because doing so does nothing to the miners.

Security Under Double Spending Attacks. In traditional transactions, because there is a centralized institution such as a bank, there is no double spending problem: each payment will be deducted from your bank account, and all details are available in the bank recording. However, in blockchain systems such as Bitcoin, there is a danger of double spending [11,22] when a transaction occurs because there is no guarantee from a centralized institution such as a bank. In our scheme, transaction packaging and publishing to the final blockchain are completed by different nodes, and miners cannot change the content in the transaction list, because the consensus is reached by the consensus group and each list has a signature of the consensus group. Miners can only put the complete transaction list into the microblock, and then put multiple microblocks into the block of the final blockchain. In this way, our scheme can resist double spending attacks.

Security Under Flash Attacks. In flash attacks [5], an attacker can obtain a large amount of computing power temporarily to launch an attack on a blockchain system. This attack can break its security assumptions in the traditional proof-of-work mechanism. But in our system, flash attacks cannot be implemented. Because the transaction list is packaged with high-reputation nodes, and the mining of microblocks does not require too much computing power, so even if the adversary can temporarily have a large amount of computing power, he can only mine more blocks, and cannot pose a threat to the transaction blockchain.

6.2   Performance Analysis

For our prototype, we built an experimental network and implemented our protocol. We deploy the nodes on client machines with Intel i7-4600U 2.70 GHz CPU, and 16 GB RAM. In the experiment we set up 1000 nodes, each node has a bandwidth of 20 Mbps.

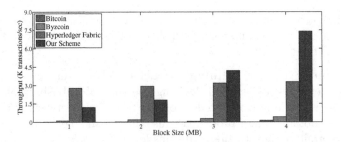

Fig. 5. Throughput of the scheme.

Figure 5 compares our scheme with Bitcoin, ByzCoin and Hyperledger Fabric [3] in terms of throughput in different block sizes of 0.5 MB, 1 MB, 2 MB and 4 MB. The Bitcoin system is the most commonly used blockchain system, and it uses a proof-of-work consensus mechanism. However, due to its throughput limitations, it is difficult to use in the field of e-commerce. In ByzCoin, the set of witnesses is dynamically formed, a new leader is elected each time a new key block is created, and the set of transactions can be committed every few seconds if a sufficient number of witnesses have collectively signed it. Hyperledger Fabric is the one with higher performance in the current blockchain system, but its degree of centralization is high, and the chain is limited to members of the alliance.

We can see that the throughput of the Bitcoin system is 2 TPS, 3 TPS, 7 TPS, 14 TPS under different block sizes, the ByzCoin system is 105 TPS, 205 TPS, 287 TPS and 428 TPS respectively. Hyperledger Fabric's throughput is approximately 2785 TPS, 2940 TPS, 3185 TPS and 3285 TPS respectively, which is a blockchain system that is advantageous in terms of throughput. In Our scheme, when the block size is 0.5 MB, the throughput is about 1200 TPS. When the block size is 1MB, its throughput is close to 2000 TPS and when the block sizes are 2 MB and 4 MB, the throughput exceeds 4000 TPS and 7000 TPS, respectively.

Through the experimental results, it can be seen that our scheme has a high throughput and is suitable for transaction scenarios.

7   Conclusion

In this paper, we design a reputation based consensus mechanism with high security and efficiency for e-commerce blockchain. In the proposed scheme, we have established a reputation mechanism to supervise the legal behavior of nodes in terms of consensus and transactions. We use the reputation chain to implement distributed storage of reputation, and enable users to publish multiple reputation evaluation information for a transaction. Besides, the scheme decouples the transaction serialization and writes it into the blockchain, adding microblocks prevents the concentration of computing power and increases fairness. The results of the security and performance analysis demonstrate that the proposed scheme can be well adapted to the application scenarios of transaction.

Acknowledgment. The authors acknowledge the support from National Key R&D Program of China under Grant No.2017YFB1400700 and National Natural Science Foundation of China under Grant No.: 61772514.

References

1. Delegated Proof-of-Stake Consensus. https://bitshares.org/technology/delegated-proof-of-stake-consensus/
2. POA Network Whitepaper (2018). https://github.com/poanetwork/wiki/wiki/POA-Network-Whitepaper
3. Androulaki, E., et al.: Hyperledger fabric: a distributed operating system for permissioned blockchains (2018)
4. Bentov, I., Pass, R., Shi, E.: Snow white: provably secure proofs of stake. IACR Cryptol. ePrint Arch. **2016**, 919 (2016)
5. Bonneau, J.: Why buy when you can rent? In: Clark, J., Meiklejohn, S., Ryan, P.Y.A., Wallach, D., Brenner, M., Rohloff, K. (eds.) FC 2016. LNCS, vol. 9604, pp. 19–26. Springer, Heidelberg (2016). https://doi.org/10.1007/978-3-662-53357-4_2
6. Chen, J., Micali, S.: Algorand (2016)
7. community, N.: NXT Whitepaper (2014)
8. Eyal, I., Sirer, E.G.: Majority is not enough: bitcoin mining is vulnerable. Commun. ACM **61**(7), 95–102 (2018)
9. Gai, F., Wang, B., Deng, W., Peng, W.: Proof of reputation: a reputation-based consensus protocol for peer-to-peer network. In: Pei, J., Manolopoulos, Y., Sadiq, S., Li, J. (eds.) DASFAA 2018. LNCS, vol. 10828, pp. 666–681. Springer, Cham (2018). https://doi.org/10.1007/978-3-319-91458-9_41
10. Jakobsson, M., Juels, A.: Proofs of work and bread pudding protocols (extended abstract). In: IFIP TC6/TC11 Joint Working Conference on Secure Information Networks: Communications & Multimedia Security (1999)
11. Karame, G., Androulaki, E., Capkun, S.: Double-spending fast payments in bitcoin. In: CCS 2012, Raleigh, NC, USA, pp. 906–917 (2012)
12. Kiayias, A., Tsiounis, Y., Yung, M.: Traceable signatures. In: Cachin, C., Camenisch, J.L. (eds.) EUROCRYPT 2004. LNCS, vol. 3027, pp. 571–589. Springer, Heidelberg (2004). https://doi.org/10.1007/978-3-540-24676-3_34
13. Kiayias, A., Russell, A., David, B., Oliynykov, R.: Ouroboros: a provably secure proof-of-stake blockchain protocol. In: Katz, J., Shacham, H. (eds.) CRYPTO 2017. LNCS, vol. 10401, pp. 357–388. Springer, Cham (2017). https://doi.org/10.1007/978-3-319-63688-7_12
14. King, S., Nadal, S.: PPCoin: peer-to-peer crypto-currency with proof-of-stake (2012)

15. Lamport, L.: The part-time parliament. ACM Trans. Comput. Syst. **16**(2), 133–169 (1998)
16. Lamport, L., Shostak, R., Pease, M.: The Byzantine generals problem (1982)
17. Li, H., Huang, H., Tan, S., Zhang, N., Fu, X., Tao, X.: A new revocable reputation evaluation system based on blockchain. IJHPCN **14**(3), 385–396 (2019)
18. Miguel, O.T.D.C.: Practical Byzantine fault tolerance. ACM Trans. Comput. Syst. **20**(4), 398–461 (2002)
19. Nakamoto, S.: Bitcoin: a peer-to-peer electronic cash system (2008, consulted)
20. Ongaro, D., Ousterhout, J.K.: In search of an understandable consensus algorithm. In: USENIX ATC 2014, Philadelphia, PA, USA, 19–20 June 2014, pp. 305–319 (2014)
21. Pass, R., Shi, E.: Hybrid consensus: efficient consensus in the permissionless model. In: DISC 2017, Vienna, Austria, pp. 39:1–39:16 (2017)
22. Rosenfeld, M.: Analysis of hashrate-based double spending. Eprint Arxiv (2014)
23. Schiller, E.M., Schwarzmann, A.A. (eds.): PODC 2017, Washington, DC, USA, 25–27 July 2017. ACM (2017)
24. Vasin, P.: BlackCoins Proof-of-Stake Protocol v2 (2017). www.blackcoin.co
25. Yu, J., Kozhaya, D., Decouchant, J., Veríssimo, P.J.E.: RepuCoin: your reputation is your power. IEEE Trans. Comput. **68**(8), 1225–1237 (2019)

A Secure and Efficient Smart Contract Execution Scheme

Zhaoxuan Li[1,2], Rui Zhang[1,2(✉)], and Pengchao Li[1]

[1] State Key Laboratory of Information Security,
Institute of Information Engineering, Chinese Academy of Sciences,
Beijing 100093, China
{lizhaoxuan,zhangrui,lipengchao}@iie.ac.cn
[2] School of Cyber Security, University of Chinese Academy of Sciences,
Beijing 100049, China

Abstract. As a core technology of the blockchain, the smart contract is receiving increasing attention. However, the frequent outbreak of smart contract security events shows that improving the security of smart contracts is essential. How to guarantee the privacy of contract execution and the correctness of calculation results at the same time is still an issue to be resolved. Using secure multi-party computation (SMPC) technology to implement smart contracts is considered to be one of the potential solutions. But in the existing SMPC based contract execution schemes, a problem has been ignored, that is, the attacker can perform the same process as the reconstructor to recover the secret, which leads to the leakage of users' privacy. Therefore, in order to solve this problem in the process of smart contract operation, an improved homomorphic encryption algorithm is proposed in this paper, which has a relatively small public key size, short ciphertext length, and high encryption efficiency. Then, a contract execution scheme integrated with SMPC and homomorphic encryption (SMPC-HE for short) is further proposed, which is able to guarantee the privacy of contract execution and the correctness of the calculation results at the same time, and also makes smart contract execution fairer. Finally, our scheme is proved secure, efficient and has low space overhead by theory and experiment results.

Keywords: Smart contract · Blockchain · Secure multi-party computation · Homomorphic encryption · Privacy protection

1 Introduction

The concept of smart contracts was put forward by Szabo [12] in 1994. However, smart contracts have not found a suitable application scenario since then. It was not until the advent of blockchain technology that smart contracts, as a core technology of it, re-entered the public's vision [7]. Because its Turing completeness can realize complex blockchain applications, it has been widely used in the Internet of Things, medical health, and other fields. But due to the immutability

© Springer Nature Switzerland AG 2020
W.-S. Ku et al. (Eds.): ICWS 2020, LNCS 12406, pp. 17–32, 2020.
https://doi.org/10.1007/978-3-030-59618-7_2

of the blockchain, contracts cannot be changed once deployed, even if there are vulnerabilities in the contract itself. In addition, the contract itself carries huge economic value, thus increasingly becomes the target of hackers. In April 2018, the US chain's token BEC caused a market value of \$900 million to almost zero due to a security breach in the code. In May 2019, Binance suffered hacking and led to the theft of more than 7,000 Bitcoins. These frequent security incidents of smart contracts show that improving the security of smart contracts is essential.

As an important part of smart contract security, operation security mainly ensures that contracts are executed more securely and accurately. Since the execution of contracts involves a large amount of users' privacy, how to ensure the privacy of contract execution and the correctness of calculation results at the same time is a question worthy of study. Using secure multi-party computation (SMPC) to implement contracts is considered one of the most potential solutions. At present, some works [2,5,8,18] have used SMPC to ensure the secure operation of contracts. But in these existing schemes, a problem has been ignored, that is, the attacker can perform the same process as the reconstructor to recover the secret, which leads to the leakage of users' privacy. All in all, these works have conducted a beneficial exploration of contract execution security, but not fundamentally solved the execution security problem of contracts.

A secure and efficient smart contract execution scheme is proposed, which consists of two parts: on-chain and off-chain. In order to efficiently complete the secure calculation in SMPC and homomorphic encryption (HE), the confirmation contract and currency settlement are performed on-chain, and operations such as key generation, data encryption, and decryption are implemented off-chain. Specifically, this paper makes the following contributions:

- Since the storage space and processing power of the blockchain are special, a more space-saving and more efficient HE algorithm is proposed. The number of encrypted plaintext bits is k, and each component of the public key is in the form of cubic. The algorithm achieves IND-CPA security. Compared with the existing HE algorithm, this algorithm has a relatively small public key size, short ciphertext length, and high encryption efficiency.
- An SMPC-HE based smart contract execution scheme is proposed, which combines SMPC with the improved HE algorithm. This scheme not only makes the user's privacy better protected during the execution of the smart contract, but also ensures the correctness of the calculation results. Moreover, this scheme greatly improves the fairness and security of the smart contract.

The rest of this paper is organized as follows: in Sect. 2, we introduce the research status of related technologies. Section 3 expounds the proposed smart contract execution scheme in detail. Section 4 gives the safety, efficiency analysis and experimental comparison. Finally, Sect. 5 summarizes the paper.

2   Related Work

2.1   Secure Multi-party Computation

In 1979, Shamir [11] proposed a threshold secret sharing scheme. Subsequently, Andrew et al. [17] proposed the concept of SMPC for the first time. In recent years, some references have proposed the use of SMPC to implement contracts. In [2], a time-dependent commitment is proposed to extend the instruction set in Bitcoin. In [5], a two-party fairness protocol supporting "claim-or-refund" was designed using the Bitcoin network and extended to the design of an SMPC protocol with penalties. The above works are mainly researches on the Bitcoin network. However, no one discussed the security implementation of commonly used contracts. Afterward, Pei et al. [8] used an SMPC without a trusted third party to improve the security of contracts, but the efficiency of this scheme has not been verified, and the multiplication cannot be performed. Yan et al. [18] proposed a contract framework based on SMPC. However, malicious users can imitate the reconstructor to collect the calculation result information to recover the secret in these schemes. Since the secret sharing of some contracts involves the users' privacy, this security loophole will lead to the leakage of users' privacy. The above works reflect the advantages of SMPC, but there are security loopholes in these works. Therefore, it is necessary to further study the application of SMPC in smart contracts.

2.2   Homomorphic Encryption

In 1978, Rivest et al. [10] proposed the concept of homomorphic encryption. In 2013, Plantard et al. [9] proposed a fully homomorphic encryption scheme based on ideal lattices. In the encryption process, as the operation of the ciphertext increases, the noise also increases. In 2010, the DGHV scheme was given in [14], which supports operations of addition and multiplication on integers. Then in [6,15,16], the plaintext space and the form of public-key were improved based on DGHV. A new homomorphic encryption method was proposed in [13], but this method is not secure. Among the existing integer-based HE algorithms, some algorithms can only encrypt one bit of data, and some encryption operations need a lot of space to store the public key. Moreover, the security of some algorithms is attributed to the partial approximate greatest common divisor problem (PACDP) [6], some algorithms are not random, and some have complex operations, which result in poor security, low efficiency and huge space overhead.

In summary, the current schemes of smart contracts based on SMPC have not been able to protect users' privacy well. Moreover, the above HE algorithm cannot meet the blockchain's demands for efficiency and space overhead. Therefore, we improved the HE algorithm and integrated it with SMPC to make the execution of smart contracts more secure and efficient.

3 SMPC-HE Based Smart Contract Execution Scheme

3.1 Overview

In this paper, a secure and efficient smart contract execution scheme is proposed from the aspect of operation security. Figure 1 describes the main framework of this scheme, which consists of on-chain and off-chain. Due to the basic properties of the public chain, the chain includes distributed ledgers, transactions and smart contracts. Moreover, confirmation contracts and currency settlements are mainly performed on-chain. In order to complete secure computing efficiently in SMPC-HE, many tasks should be implemented off-chain, such as key generation, data encryption and decryption. In addition, sensitive data should not be recorded directly on the blockchain. IPFS [1] distributed storage system is used as a chain component because of its feasibility. After the contract has been audited automatically, it will be deployed on the public chain if there are no vulnerabilities, the contract code will be stored in the off-chain database, and the contract storage location, as well as contract hash, will be stored on-chain.

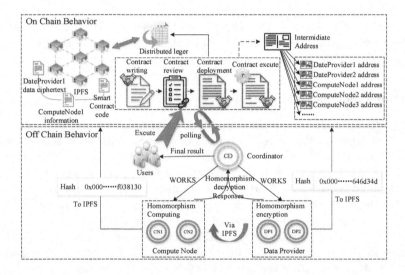

Fig. 1. The main framework of the smart contract execution scheme.

For execution security, the smart contract uses an intermediate address instead of the actual address. SMPC-HE requires each blockchain node to vote for a reliable node (called Coordinator) according to consensus, which is used to connect on-chain and off-chain message exchanges, manage and supervise the execution of smart contracts, and coordinate the relationship between nodes in the execution process. When the smart contract is called and executed by the user, the Coordinator checks whether the contract code stored off-chain is correct based on the hash value stored on-chain. Moreover, the user needs to provide the data required by the smart contract and pay for the operation gas to the Coordinator and blockchain nodes. Then the Coordinator calculates whether the user meets the conditions for calling smart contracts and whether the user's input is correct. If all conditions are met, subsequent operations will be completed. The Coordinator selects trusted computing nodes and uses SMPC-HE to process the users' data encrypted by homomorphic encryption, so as to obtain the final result and update it to the blockchain (see Sect. 3.3 for details). In fact, the Coordinator does not participate in the calculation of smart contracts, maintains zero-knowledge during the execution process and only plays the role of verification and notification. Therefore, the existence of the Coordinator does not affect the fairness of smart contracts.

3.2 The Improved Homomorphic Encryption Algorithm

The biggest advantage of homomorphic encryption is that the encrypted data can be calculated, and the result is equal to that of the same calculation using plaintext directly. This not only shortens the calculation process, but also makes it difficult to expose the unencrypted data to attackers. Due to the special storage space and processing power of the blockchain, a more space-saving and more efficient HE algorithm is needed. Therefore, we improve the existing HE algorithm to meet the needs of the blockchain. Table 1 lists the descriptions of variable symbols used in the homomorphic encryption.

Table 1. Description of variable symbols in homomorphic encryption

Symbol	Meaning
γ	The bit length of the integer in the public key
η	The bit length of the integer in the private key
ρ	The bit length of noise integer
τ	The number of integers in the public key
ρ'	The bit length of the second noise integer
k	The bit length of an integer in plaintext
λ	Safety parameters
$\lfloor z \rfloor, \lceil z \rceil, \lfloor z \rceil$	The up, down, and nearest rounding sign of the real number z $\lceil z \rceil \in [z, z+1), \lceil z \rceil \in (z-1, z], \lfloor z \rceil \in (z - 1/2, z + 1/2]$
$q_p(z)$	The quotient of the integer z divided by the integer p, $q_p(z) = \lfloor z \rceil$
$r_p(z), [z]_p$	The remainder of the integer z divided by the integer p $r_p(z) = z - q_p(z), r_p(z) \in (-p/2, p/2]$
$\tilde{O}()$	Infinite mass of the same order
$\omega()$	Infinite mass of the higher order

We construct an integer $x_{i,j,k} = x_{i,0} \cdot x_{j,1} \cdot x_{k,2} \bmod x_0$, where i,j,k satisfies $1 \le i,j,k \le \beta$ ($\beta = \sqrt[3]{\tau}$). In order to generate $\tau = \beta^3$ integers $x_{i,j,k}$, only 3β integers $x_{i,b}$ need to be stored in the public key. In other words, we use the cubic form of the public key element instead of the linear form. In this way, the number of integers in the public key is roughly reduced from τ to $3\sqrt[3]{\tau}$. Moreover, the plaintext space $\{0,1\}$ of DGHV [14] is extended to $\{0,1\}^k$ by using the operation in the number field $\{0,1\}^k$ to realize the HE algorithm of encrypting k bit plaintext $m \in \{0,1,\cdots,2^k-1\}$ into a ciphertext. The encryption algorithm consists of 4 parts: KeyGen(), Encrypt(), Decrypt(), and Evaluate().

KeyGen(1^λ): Generate a η bit random number $p \leftarrow (2\mathbb{Z}+1) \cap [2^{\eta-1}, 2^\eta)$, and use p as the private key, that is $sk = p$; for the public key pk, sample from the $D_{\gamma,\rho}(p)$ distribution to obtain $3\beta + 1$ integers $x \leftarrow D_{\gamma,\rho}(p)$, and mark the largest integer as x_0 ($x_0 \in 2\mathbb{Z}+1, r_p(x_0) \in 2\mathbb{Z}$, if x_0 is not satisfied, then recalculate), and the remaining integers are respectively labeled $x_{i,b}$ ($1 \le i \le \beta$ and $b \in \{0,1,2\}$). The resulting public key is $pk = \langle x_0, x_{1,0}, x_{1,1}, x_{1,2}, \cdots, x_{\beta,0}, x_{\beta,1}, x_{\beta,2} \rangle$.

$$D_{\gamma,\rho}(p) = \left\{ q \leftarrow \mathbb{Z} \cap [0, 2^\gamma/p), r \leftarrow \mathbb{Z} \cap (-2^\rho, 2^\rho) : Output\, x = pq + 2^k r \right\} \quad (1)$$

Encrypt($pk, m \in \{0,1\}^k$): Randomly select three subsets $S_i, S_j, S_k \in \{1,2,\cdots,\beta\}$, and randomly generate a vector $b = \langle b_{i,j,k} \rangle$ with dimension $len(S_i) \cdot len(S_j) \cdot len(S_k)$ ($len(S_i)$ denotes the number of elements in the set (S_i) and the vector coefficient $b_{i,j,k} \in \mathbb{Z} \cap [0, 2^\alpha)$. Then a random number $r \in \mathbb{Z} \cap (-2^{\rho'}, 2^{\rho'})$ is generated. The output ciphertext is:

$$c \leftarrow \left[m + 2^k r + 2 \sum_{i \in S_i, j \in S_j, k \in S_k} b_{i,j,k} \cdot x_{i,0} \cdot x_{j,1} \cdot x_{k,2} \right]_{x_0} \quad (2)$$

Decrypt(sk, c): Enter the private key $sk = p$ and the ciphertext c. Then decrypt to get the plaintext m', that is:

$$m' \leftarrow (c \bmod p) \bmod 2^k \quad (3)$$

In the decryption phase, set the value of $c \bmod p$ in the range $(-p/2, p/2]$, then $c \bmod p = c - p \cdot \lfloor c/p \rceil$. Since p is an odd number, the size of the integer can be reduced in the operation by $m' \leftarrow [[c]_{2^k} - [p \cdot \lfloor c/p \rceil]_{2^k}]_{2^k}$ and obtain m'.

Evaluate($pk, C, c_1, \cdots, c_t$): Given a t bit binary gate circuit C and t ciphertexts c_i, the addition and multiplication gate circuits in C are used to perform process the ciphertext on integers to obtain an integer as the new ciphertext $c^* = $ Evaluate$(pk, C, c_1, \cdots, c_t)$ and satisfies Decrypt$(sk, c^*) = C(m_1, \cdots, m_t)$.

Lemma 1. *Assume* $(sk, pk) \leftarrow KeyGen(1^\lambda)$, $m \in \{0,1\}^k$, $c \leftarrow Encrypt(pk, m)$, *then the ciphertext c in this HE algorithm has the following form for some integers a and b:* $c = m + ap + 2^k b$, *and satisfies* $|m + 2^k b| \le \tau 2^{\rho'+3k+3}$.

Proof. For the convenience of explanation, let $\phi = i \in S_i, j \in S_j, k \in S_k$. Known from this HE algorithm: $c \leftarrow [m + 2^k r + 2\sum_\phi b_{i,j,k} x_{i,0} x_{j,1} x_{k,2}]_{x_0}$. Because of $|x_0| \geqslant |x_{i,b}|$ for any $1 \leqslant i \leqslant \beta$ and $b \in \{0,1,2\}$, there is an integer n such that the ciphertext $c = m + 2^k r + 2\sum_\phi b_{i,j,k} x_{i,0} x_{j,1} x_{k,2} + n x_0$, where $|n| \leqslant \tau$. Since x_0 and $x_{i,b}$ are both generated by the formula $D_{\gamma,\rho}(p)$, it is assumed that $x_0 = pq_0 + 2^k r_0$ and $x_{i,b} = pq_{i,b} + 2^k r_{i,b}$. Further $c = m + 2^k r + 2\sum_\phi b_{i,j,k}(pq_{i,0} + 2^k r_{i,0})(pq_{j,1} + 2^k r_{j,1})(pq_{k,2} + 2^k r_{k,2}) + n(pq_0 + 2^k r_0) = m + p(nq_0 + 2\sum_\phi b_{i,j,k}(p^2 q_{i,0} q_{j,1} q_{k,2} + \cdots + 2^{2k} q_{k,2} r_{i,0} r_{j,1})) + 2^k(r + n r_0 + 2^{2k+1}\sum_\phi b_{i,j,k} r_{i,0} r_{j,1} r_{k,2})$. It can be seen that the ciphertext c has the form $c = m + ap + 2^k b$ for some integers a and b. The inequality can be obtained from $\rho' \geqslant 3\rho + \alpha$: $|m + 2^k(r + n r_0 + 2^{2k+1}\sum_\phi b_{i,j,k} r_{i,0} r_{j,1} r_{k,2})| \leqslant 2^{\rho'+k} + \tau 2^{\rho+k} + \tau 2^{3k+3\rho+\alpha+1} \leqslant (2\tau + 1) 2^{\rho'+3k+1} \leqslant \tau 2^{\rho'+3k+3}$. Therefore, the above lemma has been proved.

Theorem 1. *The improved HE algorithm is correct.*

Proof. Let C be an operational circuit with t bit inputs. $c_i = \text{Encrypt}(pk, m_i)$ and $c \bmod p = C(c_1, \cdots, c_t) \bmod p = C(c_1 \bmod p, \cdots, c_t \bmod p) \bmod p$ are known. From Lemma 1, we can know $m_i = [c_i \bmod p]_{2^k}, i \in [1, t]$. As well as, $|C(c_1 \bmod p, \cdots, c_t \bmod p)| \leqslant 2^{(\eta-4)(k-1)} \leqslant (p/8)^{k-1} \leqslant p$ and $C([c_1 \bmod p]_{2^k}, \cdots, [c_t \bmod p]_{2^k}) \leqslant 2^k$ can be obtained from the definition and nature of operable circuit, so: $c \bmod p = C(c_1 \bmod p, \cdots, c_t \bmod p)$. Further $[c \bmod p]_{2^k} = C([c_1 \bmod p]_{2^k}, \cdots, [c_t \bmod p]_{2^k}) \bmod 2^k = C([c_1 \bmod p]_{2^k}, \cdots, [c_t \bmod p]_{2^k}) = C(m_1, \cdots, m_t)$. So the algorithm is correct.

Theorem 2. *The algorithm is homomorphic with addition and multiplicative.*

Proof. Assuming two messages m_0 and m_1, encrypting them separately can get: $c_0 = [m_0 + 2^k r_0 + 2\sum_{i \in S_i, j \in S_j, k \in S_k} b_{i,j,k} x_{i,0} x_{j,1} x_{k,2}]_{x_0}$, $c_1 = [m_1 + 2^k r_1 + 2\sum_{i \in S_i, j \in S_j, k \in S_k} b_{i,j,k} x_{i,0} x_{j,1} x_{k,2}]_{x_1}$. From Lemma 1, there are integers A_0, B_0, A_1, and B_1, which make: $c_0 = m_0 + pA_0 + 2^k B_0$ and $c_1 = m_1 + pA_1 + 2^k B_1$. So we can get $c_0 + c_1 = m_0 + m_1 + p(A_0 + A_1) + 2^k(B_0 + B_1)$ and $c_0 \times c_1 = m_0 m_1 + p(m_0 A_1 + m_1 A_0 + pA_0 A_1 + 2^k A_0 B_1 + 2^k A_1 B_0) + 2^k(m_0 B_1 + m_1 B_0 + 2^k B_0 B_1)$.

Decrypting the ciphertext using the decryption algorithm, we can get: $m_+ = \text{Decrypt}(sk, c_0 + c_1) = [(c_0 + c_1) \bmod p]_{2^k} = [m_0 + m_1 + 2^k(B_0 + B_1)]_{2^k} = m_0 + m_1$ and $m_\times = \text{Decrypt}(sk, c_0 \times c_1) = [(c_0 \times c_1) \bmod p]_{2^k} = [m_0 \times m_1 + 2^k(m_0 B_1 + m_1 B_0 + 2^k B_0 B_1)]_{2^k} = m_0 \times m_1$. Therefore, the algorithm satisfies the homomorphism of addition and multiplicative.

3.3   The Operation Process of Contracts Integrated with SMPC-HE

SMPC-HE. The SMPC is n participants $P_1, \cdots, P_n$, which need to perform a certain computing task together, expressed as $F(x_1, \cdots, x_n) = (y_1, y_2, \cdots, y_n)$. Each party P_i can only get its own input x_i, and can only calculate its own output y_i. If the total number of participants is n, the number of honest parameter parties is t. The condition for SMPC to provide complete, secure, and credible results is $t \geqslant 2n/3$. In SMPC-HE, improved homomorphic encryption, SMPC, and blockchain are integrated. The SMPC has the characteristics of input privacy, calculation accuracy, etc., which can keep the data both private and usable. Moreover, during the execution of SMPC, the computing tasks of each party are consistent. This feature can guarantee the fairness of smart contracts. In SMPC, although the sub-secret does not reveal the original secret, once the attacker obtains t sub-secrets, the original secret can be reconstructed. The introduction of homomorphic encryption allows calculations to be performed on the basis of ciphertext, and even if sub-secrets are reconstructed, the original secrets are not revealed. Therefore, SMPC-HE improves the security of contracts.

Theorem 3. *The Shamir secret sharing scheme has additive homomorphism and restricted multiplicative homomorphism. Specifically, if multiple (t, n) secret sharing schemes are used to share multiple secret values, then the fragmented sum of different secret values is the fragment to the sum of corresponding secret values. If d (t, n) secret sharing schemes share multiple secret values. Then if and only if $d(t-1) \leqslant n-1$, the fragmented product of these secret values is still the fragment to the product of corresponding secret values.*

Proof. Due to the limited space, the proof process of the additive and multiplicative homomorphism is detailed in [3,4].

Assume that there are n participants. The SMPC-HE scheme is illustrated by calculating the sum and difference between a and b. Figure 2 shows the flow of the overall data. The detailed process of SMPC-HE's addition and subtraction operation is mainly divided into the following three phases:

(1) *Verifiable secret sharing.* Homomorphic encryption and (t, n) secret sharing schemes are used to complete secret sharing. In view of the users' privacy involved in some secrets, the reason for introducing improved homomorphic encryption is to prevent malicious users from obtaining sub-secrets from multiple computing nodes to reconstruct secrets, leading to privacy leaks. Moreover, SMPC is performed on the basis of ciphertext, so the encryption scheme must be homomorphic. Therefore, an improved HE algorithm with high efficiency and small space consumption is introduced. The specific process of secret sharing is as follows: for a given secret a, use the homomorphic encryption public key of the trusted party Coordinator to homomorphic encrypt the secret a to obtain the ciphertext $d = \text{Encrypt}(pk, a)$. Generate n shared values $d \to \{d_1, \cdots, d_n\}$ from the ciphertext d, where $n \geqslant 1$. The process of generating shared values is: set $r_0 = d$, and $t-1$ random numbers $(r_1, r_2, \cdots, r_{t-1})$ in F_p are randomly selected to construct polynomial equation $f_d(x) = \sum_{i=0}^{t-1} r_i x^i$. For the trusted computing node P_i (where $i \in [1, n]$) with identity θ_i, the shared sub-secret is (θ_i, d_i), where $d_i = f_d(\theta_i)$; similarly, the given secret b is homomorphic encrypted to obtain the ciphertext $e = \text{Encrypt}(pk, b)$. Generate n shared values $e \to \{e_1, \cdots, e_n\}$ from the ciphertext e, where $n \geqslant 1$. Then the shared sub-secret is (θ_i, e_i), $i \in [1, n]$.

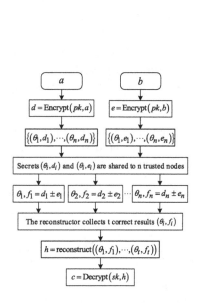

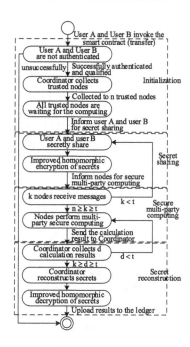

Fig. 2. Data flow of SMPC-HE addition and subtraction

Fig. 3. The time sequence and state transition of the smart contract.

In order to make secret sharing and reconstruction more secure, a verifiable secret sharing mechanism is further introduced to verify the correctness of the secret. The security of the verifiable secret sharing mechanism can be reduced to the discrete logarithm problem. Take a p order generator g of the multiplication group $\mathbb{Z}_q^*$ to obtain a cyclic subgroup $\langle g \rangle$ (where $p \mid (q-1)$), and find the sets $G = \{g_i = g^{r_i} \bmod q\}_{i=0}^{t-1} \cup \{g\}$ and $K = \{k_i = g^{r_i'} \bmod q\}_{i=0}^{t-1} \cup \{g\}$. Then the set G and the set K need to be sent to the trusted computing nodes.

(2) *Secure multi-party computing.* Each trusted computing node P_i $(i \in [1, n])$ receives the shared sub-secrets (θ_i, d_i), (θ_i, e_i) and sets G, K. If the equations $g^{d_i} = g^d \prod_{j=1}^{t-1} g^{r_j \theta_i{}^j} = \prod_{j=0}^{t-1} g_j^{\theta_i{}^j}$ and $g^{e_i} = g^e \prod_{j=1}^{t-1} g^{r_j' \theta_i{}^j} = \prod_{j=0}^{t-1} k_j^{\theta_i{}^j}$ are satisfied, the sub-secrets are successfully received. The number of nodes successfully receiving sub-secrets must be not less than t. Then perform their own security calculations, which are mainly divided into addition and multiplication. Of course, it can also perform subtract and divide, that is, add a negative number and multiply by the inverse of a number. Each P_i sends the result (θ_i, f_i) to the reconstructing party after the security calculation is completed.

(3) *Secret reconstruction.* The reconstruction party chooses to receive t correct results (θ_i, f_i) from n trusted computing nodes P_i $(i \in [1, n])$. If the equation $g^{f_i} = g^d g^e \prod_{j=1}^{t-1} (g^{r_j \theta_i{}^j} \cdot g^{r_j' \theta_i{}^j}) = \prod_{j=0}^{t-1} (g_j k_j)^{\theta_i{}^j}$ is satisfied, the

calculation result is correct. Then use the *Lagrange* interpolation method to recover the total calculation result $h' = \text{reconstruct}\,((\theta_1, f_1), \cdots, (\theta_t, f_t))$. The recovery process is as follows: at first, recover the polynomial equation $f_h(x) = \sum_{j=1}^{t} y_j L_j(x) = \sum_{j=1}^{t} (f_j \prod_{i=1, i \neq j}^{t} \frac{x - \theta_i}{\theta_j - \theta_i})$. Then recover the results $h' = f_h(0) = \sum_{j=1}^{t} (f_j \cdot \prod_{i=1, i \neq j}^{t} \frac{-\theta_i}{\theta_j - \theta_i})$, $h' = h$. Since h' is ciphertext, the homomorphic encryption private key of the trusted party Coordinator needs to be used to homomorphic decrypt the ciphertext h' to obtain the final calculation result $c = \text{Decrypt}\,(pk, h')$.

Because of the multiplicative homomorphism of homomorphic encryption and the restricted multiplicative homomorphism of (t, n) threshold secret sharing. This scheme can also perform multiplication and division operations, and further perform mixed operations including addition, subtraction, multiplication and division. The relevant calculation method is similar to the above, so it will not be re-explained. Therefore, this scheme can be used to solve blockchain-related calculations safely and efficiently. For example, the transaction cost of the smart contract is calculated by the formula $Gas = Gas\,Limit \times Gas\,Price$.

The Operation Process of Contracts Integrated with SMPC-HE. Figure 3 shows the time sequence and state transition of the smart contract at runtime. In order to describe the process intuitively and clearly, the transfer contract is used as an example to explain. Due to the limited space, it mainly describes the operations between the two parties of the contract execution. The correctness of the scheme depends on the addition and restricted multiplicative homomorphism of the homomorphic encryption and (t, n) threshold secret sharing scheme, so the calculation result of the scheme is correct. The detailed operation process of the smart contract is divided into the following four phases:

Initialization: In this phase, the following three operations are performed:

(1) Authenticate the two parties of the smart contract execution and review their conditions. The standard challenge-response protocol is used to authenticate the two parties of the smart contract execution. The specific process of authenticating the identity of user A is as follows: the authenticator Coordinator uses the public key derived from secp256kl elliptic curve to obtain the address of user A and sends a challenge to user A. User A uses the private key and the elliptic curve digital signature algorithm to sign the challenge and sends the response to the Coordinator. The Coordinator uses the public key of user A to authenticate the received response. The same method is used to authenticate the identity of user B. After identity authentication, check whether both parties meet the calling conditions of the contract.

(2) The Coordinator collects trusted nodes. The number of nodes collected m is not a fixed value, and m is changed according to actual needs. But m should satisfy $m \geq n \geq t$, and it is better to leave some redundant nodes. After the Coordinator collects the nodes, it stores the information of the computing nodes in the "trustednodes" file of the public storage area.

(3) Initialize the execution environment and data. The Coordinator connects the "trustednodes" file to read the trusted node information in the list and starts the daemon processes of n computing nodes. Then the threshold value $t = \lfloor (2 \cdot num)/3 \rfloor$ in subsequent secret sharing is calculated according to the number of participating computing nodes.

Secret Sharing: The Coordinator informs the parties of the smart contract execution to enter the secret-sharing phase. Take the transfer contract as an example, both parties of the contract execution perform the opposite operation. For user A, the operation performed is the payment, that is, the account balance of user A subtracts the transfer amount; for user B, the operation performed is the gathering, that is, the account balance of user B plus the transfer amount. In this phase, the following operations are required:

(1) User A prepares his account balance $AValue$ and the transfer amount Q. Use the homomorphic encryption public key of the Coordinator and the improved HE algorithm to encrypt the information.
(2) User A uses the (t, n) threshold secret sharing scheme to divide the encrypted secret into n sub-secrets and shares the sub-secrets to n nodes. Note that at least t nodes must be guaranteed to receive the sub-secret successfully.
(3) User B prepares his account balance $BValue$ and transfer amount Q, then shares them secretly to n nodes after encryption.
(4) When the secret-sharing is finished, the Coordinator will inform all participating nodes to perform secure multi-party calculations. However, if the number of nodes that successfully receive the sub-secret is less than t, the Coordinator will inform user A and user B to restart secret sharing.

Secure Multi-party Computing: The computing node can only obtain the secrets that need to be calculated by itself. This phase requires two operations:

(1) The trusted computing node invokes a secure computing algorithm to complete the computing process. When the computing task of this node is completed, the node will send the result to the reconstruction party Coordinator.
(2) The reconstruction party Coordinator verifies the results received. If the number of correct calculation results is less than t, the Coordinator will inform the trusted computing nodes to recalculate the secrets.

Secret Reconstruction: In this phase, the following operations are performed:

(1) After receiving the t correct calculation results, the reconstruction party Coordinator reconstructs the secret using the *Lagrange* interpolation method.
(2) Since the reconstructed secret is a ciphertext, the reconstruction party Coordinator needs to use its own private key to homomorphic decrypt the ciphertext to obtain the final calculation result.
(3) The reconstruction party Coordinator uploads the calculation results to the blockchain to update the balance status of user A and user B. In this process, it also involves mining, consensus, and other processes. These processes are the regular operations of the blockchain, which are not described here.

4   Analysis and Evaluation

4.1   The Analysis of the HE Algorithm

Theorem 4. *Fix the parameters $(\rho, \rho', \eta, \gamma, \tau)$ as in the improved HE algorithm (all polynomial in the security parameter λ). Any attack $\mathcal{A}$ with advantage ε on the encryption algorithm can be converted into an algorithm $\mathcal{B}$ for solving (ρ, η, γ)-approximate-gcd with success probability at least $\varepsilon/2$. The running time of $\mathcal{B}$ is polynomial in the running time of $\mathcal{A}$, and in λ and $1/\varepsilon$.*

Proof. This theorem can be proved by two algorithms Subroutine Lear-LSB(z, pk) and *Binary GCD*, the specific proof process can be referred to the proof of Theorem 2 in [14].

The security of this HE algorithm can be reduced to the approximate greatest common divisor problem (ACDP) [14]. The currently known attack schemes for the ACDP include violent attacks on the remainder, etc. [16]. However, the security of the encryption algorithm can be guaranteed by setting reasonable security parameters. According to Theorem 4, this algorithm achieves IND-CPA. Up to now, ACDP can't be cracked, so the algorithm is relatively secure. Moreover, ACDP is more difficult than PACDP [6]. That is to say, if the parameters are consistent, the algorithm of security to ACDP is more secure than the algorithm of security to PACDP.

Table 2. Comparison of several homomorphic schemes

Scheme	Plaintext bits/bit	Public key form	Public key size	Private key size	Encryption efficiency	Security
DGHV [14]	1	Linear	$\tilde{O}\left(\lambda^{10}\right)$	$\tilde{O}\left(\lambda^{2}\right)$	1 bit/time	ACDP
[16]	k	Linear	$\tilde{O}\left(\lambda^{10}\right)$	$\tilde{O}\left(\lambda^{2}\right)$	k bit/time	ACDP
[6]	1	Quadratic	$\tilde{O}\left(\lambda^{7}\right)$	$\tilde{O}\left(\lambda^{2}\right)$	1 bit/time	PACDP
[13]	1	Linear	$\tilde{O}\left(\lambda^{5}\right)$	$\tilde{O}\left(\lambda^{2}\right)$	1 bit/time	PACDP
[15]	k	Cubic	$\tilde{O}\left(\lambda^{6}\right)$	$\tilde{O}\left(\lambda^{2}\right)$	k bit/time	PACDP
This scheme	k	Cubic	$\tilde{O}\left(\lambda^{6}\right)$	$\tilde{O}\left(\lambda^{2}\right)$	k bit/time	ACDP

It can be seen from Table 2 that DGHV has the largest key overhead, the lowest encryption efficiency, and the length of the ciphertext generated by encrypting the plaintext of the same length is largest. The scheme in [16] changed the number of plaintext bits on the basis of DGHV and improved the encryption efficiency, but the public key size did not be changed. In [6], the scheme reduced the size of the public key to a certain extent, but the encryption efficiency did not be improved. The security of schemes in [13,15] can be reduced to PACDP, while the security of the scheme in this paper is reduced to ACDP. So the scheme in this paper is more secure. The public key of the scheme in [13] is $pk = (N, x)$. Although the length of the public key has changed to $\tilde{O}\left(\lambda^{5}\right)$, the randomness of the ciphertext has been reduced, which has greatly reduced the security.

The size of the public key in this scheme is $\tilde{O}\left(\lambda^6\right)$, which is similar to $\tilde{O}\left(\lambda^5\right)$, but the security and the encryption efficiency are greatly improved. In addition, compared with the scheme in [15], $2^k \sum_{1 \leqslant i,j,k \leqslant \beta} b_{i,j,k} x_{i,0} x_{j,1} x_{k,2}$ in the encryption process is replaced with $2 \sum_{i \in S_i, j \in S_j, k \in S_k} b_{i,j,k} x_{i,0} x_{j,1} x_{k,2}$. The randomness of the ciphertext is increased and the amount of encryption is reduced, so that the security and the encryption efficiency are improved. To sum up, compared with the existing HE algorithms, a relatively small public key size, short ciphertext length, and high encryption efficiency are obtained in this algorithm, thereby saving system encryption time. Moreover, this algorithm guarantees correctness, homomorphism, and security. In this algorithm, we find that as the form of component increases, the size of the public key will slowly decrease to $\tilde{O}\left(\lambda^5\right)$, but the calculation efficiency will decrease sharply. So considering the efficiency and space overhead, the component of the public key takes the form of cubic, and no higher form is used.

4.2 The Analysis of the Smart Contract Execution Scheme

Security. The smart contract execution scheme has significantly improved the security of smart contracts. First, during the process of contract execution, an HE algorithm is improved, which achieves IND-CPA security. The description of the relevant certification and other advantages of this algorithm are referred to Sect. 4.1. Second, SMPC-HE integrates homomorphic encryption, SMPC, and blockchain to enable contracts to complete secure computing efficiently and securely. Moreover, verifiable secret sharing is introduced in SMPC-HE, that is, the shared secret and calculation results can be verified to ensure the correctness. Therefore, the scheme guarantees the privacy of contract execution and the correctness of the calculation results at the same time. Third, the Coordinator maintains zero-knowledge throughout the process in this scheme, does not participate in calculations, and only plays the role of verification and notification. Therefore, the existence of the Coordinator does not affect the fairness of smart contracts. In other words, the contract execution does not depend on trusted third parties in this scheme. Fourth, combined with SMPC-HE, this scheme allows users to control their own data information independently and have ownership of the data. Users can agree and revoke the access rights of other users to their data. Finally, the correctness and homomorphism of the improved HE algorithm and verifiable threshold secret sharing scheme are shown in Theorem 1, 2, 3. So the scheme is feasible and maintains a high level of security.

Efficiency. The scheme adopts the on-chain and off-chain collaboration methods and makes full use of computing resources to improve operation efficiency under the condition of ensuring the security of users' privacy. Moreover, compared with the existing HE algorithms, a relatively small public key size, short ciphertext length, and high encryption efficiency are obtained in the improved HE algorithm. Therefore, the space overhead is reduced, the encryption time is saved, and the overall efficiency of the scheme is greatly improved in SMPC-HE. The larger the amount of data, the more obvious the advantages of the scheme.

4.3   The Comparative Analysis of Experiments

We use python language to implement the scheme proposed in this paper, and use the HE algorithm proposed in [6,14–16] and this paper respectively in SMPC-HE. The running environment of the program is Windows 10, Intel (R) Core (TM) i5-4200H 2.80 GHz and 12 GB RAM. In the experiment, the parameters λ and k are set to 3 and 4 respectively, and the message transmission time between nodes in the blockchain is 1 ms. Figures 4 and 5 show the changes of running time with the increase of the number of secret sharing nodes and the size of operation data in SMPC-HE based on five different HE algorithms and the contract execution scheme [18] that only uses SMPC. It can be seen from Fig. 4 that SMPC-HE based on the HE algorithm proposed in this paper consumes the least time. In addition, through the comparison between this scheme and the scheme in [18], it can be seen that the time difference between the two schemes is within 1ms, that is, the homomorphic encryption has little effect on the overall running time. Furthermore, as the number of shared nodes increases, the time for homomorphic encryption can be ignored gradually.

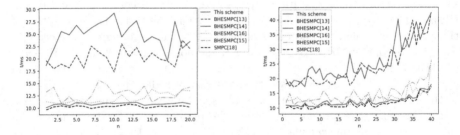

Fig. 4. Graph of sharing node number and scheme runtime.

It can be seen from Fig. 5 that SMPC-HE based on the HE algorithm proposed in this paper consumes the least time, and with the increase of data size, the time increases the slowest. In addition, through the comparison between this scheme and the scheme in [18], it can be seen that the time difference between the two schemes is about 1 ms (millisecond level) as the size of data increases, that is, the homomorphic encryption has little effect on the overall running time. Therefore, the scheme that incorporates the HE algorithm proposed in this paper has higher execution efficiency, and the time of homomorphic encryption is basically negligible compared with the time of secure multi-party calculation.

Figures 6 and 7 show the changes of memory overhead with the increase of running time and the number of secret sharing nodes in different schemes. Since the memory overhead needs to be recorded, the running time (the sum of the running time and the memory recording time, $n = 40$) in Fig. 6 is extended compared to the time in Fig. 4. It can be seen from Fig. 6 that the memory overhead in all schemes increases gradually over time, and the maximum overhead is almost the same. It can be seen from Fig. 7 that the maximum memory overhead

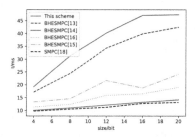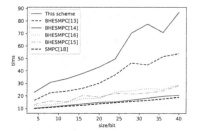

Fig. 5. Graph of calculated size and scheme runtime.

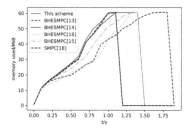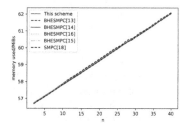

Fig. 6. Graph of runtime and memory overhead.

Fig. 7. Graph of sharing node number and maximum memory overhead.

of all schemes increases with the number of secret sharing nodes, and the maximum overhead is almost the same, so the memory overhead of the homomorphic encryption is basically negligible. Moreover, this scheme also guarantees a higher level of security. In summary, the performance of this scheme is superior in terms of security, time overhead, and space overhead.

5   Conclusion

In this paper, we propose a secure and efficient smart contract execution scheme to solve the security issues of smart contract operation in the blockchain. The scheme and its advantages are as follows: (1) An improved HE algorithm is proposed, and the correctness and homomorphism of the algorithm are proved. Compared with the existing HE algorithm, a relatively small public key size, short ciphertext length, and high encryption efficiency are obtained. (2) An SMPC-HE based smart contract execution scheme is proposed, and the correctness of the scheme is proved. This scheme not only guarantees the privacy of contract execution and the correctness of calculation results at the same time, but also makes the smart contract execution fairer. Finally, it is proved by theory and experiment that the smart contract in this scheme can meet the requirements of execution efficiency, and achieve higher security and low space overhead.

Acknowledgement. The authors acknowledge the support from National Key R&D Program of China under Grant No. 2017YFB1400700, National Natural Science Foundation of China under Grant No. 61772514 and Beijing Municipal Science & Technology Commission No. Z191100007119006.

References

1. Ali, M.S., et al.: IoT data privacy via blockchains and IPFS. In: Proceedings of the Seventh International Conference on the Internet of Things, pp. 1–7 (2017)
2. Andrychowicz, M., et al.: Secure multiparty computations on bitcoin. In: 2014 IEEE Symposium on Security and Privacy, pp. 443–458. IEEE (2014)
3. Barkol, O., Ishai, Y., Weinreb, E.: On d-multiplicative secret sharing. J. Cryptol. **23**(4), 580–593 (2010). https://doi.org/10.1007/s00145-010-9056-z
4. Benaloh, J.C.: Secret sharing homomorphisms: keeping shares of a secret secret (extended abstract). In: Odlyzko, A.M. (ed.) CRYPTO 1986. LNCS, vol. 263, pp. 251–260. Springer, Heidelberg (1987). https://doi.org/10.1007/3-540-47721-7_19
5. Bentov, I., Kumaresan, R.: How to use bitcoin to design fair protocols. In: Garay, J.A., Gennaro, R. (eds.) CRYPTO 2014. LNCS, vol. 8617, pp. 421–439. Springer, Heidelberg (2014). https://doi.org/10.1007/978-3-662-44381-1_24
6. Coron, J.-S., Mandal, A., Naccache, D., Tibouchi, M.: Fully homomorphic encryption over the integers with shorter public keys. In: Rogaway, P. (ed.) CRYPTO 2011. LNCS, vol. 6841, pp. 487–504. Springer, Heidelberg (2011). https://doi.org/10.1007/978-3-642-22792-9_28
7. Crosby, M., Pattanayak, P., Verma, S., et al.: Blockchain technology: beyond bitcoin. Appl. Innov. **2**(6–10), 71 (2016)
8. Pei, X., Sun, L., Li, X., et al.: Smart contract based multi-party computation with privacy preserving and settlement addressed. In: 2018 Second World Conference on Smart Trends in Systems, Security and Sustainability, pp. 133–139. IEEE (2018)
9. Plantard, T., et al.: Fully homomorphic encryption using hidden ideal lattice. IEEE Trans. Inf. Forensics Secur. **8**(12), 2127–2137 (2013)
10. Rivest, R.L., Adleman, L., Dertouzos, M.L., et al.: On data banks and privacy homomorphisms. Found. Secure Comput. **4**(11), 169–180 (1978)
11. Shamir, A.: How to share a secret. Commun. ACM **22**(11), 612–613 (1979)
12. Szabo, N.: Smart contracts. Virtual School (1994)
13. Tang, D., Zhu, S., Cao, Y.: Faster fully homomorphic encryption scheme over integer. Comput. Eng. Appl. **48**(28), 117–122 (2012)
14. Van Dijk, M., Gentry, C., Halevi, S., Vaikuntanathan, V.: Fully homomorphic encryption over the integers. In: Gilbert, H. (ed.) EUROCRYPT 2010. LNCS, vol. 6110, pp. 24–43. Springer, Heidelberg (2010). https://doi.org/10.1007/978-3-642-13190-5_2
15. Wang, T., Ma, W., Luo, W.: Information sharing and secure multi-party computing model based on blockchain. Comput. Sci. **46**(9), 162–168 (2019)
16. Xie, X.: An effectively fully homomorphic encryption over the integers. Master's thesis, Shandong University (2014)
17. Yao, A.C.: Protocols for secure computations. In: 23rd Annual Symposium on Foundations of Computer Science, SFCS 1982, pp. 160–164. IEEE (1982)
18. Zhu, Y., et al.: Smart contract execution system over blockchain based on secure multi-party computation. J. Cryptol. Res. **6**(2), 246–257 (2019)

A Stochastic-Performance-Distribution-Based Approach to Cloud Workflow Scheduling with Fluctuating Performance

Yi Pan[1], Xiaoning Sun[1], Yunni Xia[1(✉)], Peng Chen[2(✉)], Shanchen Pang[3], Xiaobo Li[4], and Yong Ma[5]

[1] College of Computer Science, Chongqing University, Chongqing 400044, China
xiayunni@hotmail.com
[2] School of Computer and Software Engineering,
Xihua University, Chengdu 610039, China
chenpeng@gkgb.com
[3] School of Computer and Communication Engineering,
China University of Petroleum, Qingdao 266580, China
[4] Chongqing Animal Husbandry Techniques Extension Center,
Chongqing 401121, China
[5] College of Computer Information Engineering,
Jiangxi Normal University, Nanchang 330022, China

Abstract. The cloud computing paradigm is characterized by the ability to provide flexible provisioning patterns for computing resources and on-demand common services. As a result, building business processes and workflow-based applications on cloud computing platforms is becoming increasingly popular. However, since real-world cloud services are often affected by real-time performance changes or fluctuations, it is difficult to guarantee the cost-effectiveness and quality-of-service (Qos) of cloud-based workflows at real time. In this work, we consider that workflows, in terms of Directed Acyclic Graphs (DAGs), to be supported by decentralized cloud infrastructures are with time-varying performance and aim at reducing the monetary cost of workflows with the completion-time constraint to be satisfied. We tackle the performance-fluctuation workflow scheduling problem by incorporating a stochastic-performance-distribution-based framework for estimation and optimization of workflow critical paths. The proposed method dynamically generates the workflow scheduling plan according to the accumulated stochastic distributions of tasks. In order to prove the effectiveness of our proposed method, we conducted a large number of experimental case studies on real third-party commercial clouds and showed that our method was significantly better than the existing method.

Keywords: Cloud computing · Scheduling · Workflow · Performance

Y. Pan and X. Sun—Contribute equally to this article and should be considered co-first authors.

© Springer Nature Switzerland AG 2020
W.-S. Ku et al. (Eds.): ICWS 2020, LNCS 12406, pp. 33–48, 2020.
https://doi.org/10.1007/978-3-030-59618-7_3

1 Introduction

The cloud computing architectures and services is evolving as the main-stream solution to building elastic and agile IT services and applications [1,2]. It offers platforms and reusable components for building complex applications with improved cost-effectiveness than earlier solutions, e.g., grids. The cloud management logic allocates only the required resources to the cloud users so that the system-level utilization and operational cost of resources actually used can be saved [3–5]. Based on the multiple performance criteria resource management and provisioning patterns, cloud services are delivered at different levels: infrastructure clouds (IaaS), platform clouds (PaaS), and software clouds (SaaS). The IaaS one provides resources in the form of virtual machine (VM) instances created in a data center or server node. Due to this distinctive feature, the IaaS cloud service are well recognized to be powerful and effective in supporting workflow-based applications [6].

The performance-oriented scheduling issues of cloud workflows is given considerable attention [7–9]. A widely-believed difficulty for promising user-perceived QoS is that real-world cloud infrastructures are with usually with time-varying performance and quality. Schad et al. [10] demonstrated that IaaS cloud performance in the Amazon EC2 could decrease by 23% when cloud nodes are fully loaded. Jakson et al. [11] showed that VM response time can fluctuate by 30–65% when communication delay is unstable. Such instability and performance fluctuations strongly impact the user-end QoS of business processes and applications deployed over infrastructural cloud services and probably cause violations of Service-Level-Agreement (SLA) [12]. Moreover, extra operational cost may be required to conduct fault-handling tasks or needed to counter performance degradations when SLA requirements are not met.

To face the above difficulty, we introduce a novel method to tackle the performance-fluctuation workflow scheduling problem by incorporating a stochastic-performance-distribution-based framework for estimation and optimization of workflow critical paths. The proposed method is able to estimate the accumulated stochastic distributions of workflow tasks at real-time and appropriately yield the task-VM mappings with the objective of cost reduction by optimizing the execution durations of the critical paths with the workflow completion time threshold. Case studies up on real-world commercial cloud services and multiple workflow templates demonstrate that our method beat its peers in terms of multiple performance criteria.

The main contributions of this article are as follows:

- We proposed a novel method to tackle the performance-fluctuation workflow scheduling problem by incorporating a stochastic-performance-distribution-based framework for estimation and optimization of workflow critical paths.
- Proposed method modeling time-varying performance of cloud resources and generating cost-effective scheduling plans to reduce monetary cost while following constraints of Service-Level-Agreement.

– Scheduling policy can continuously optimize the scheduling scheme in real time according to the actual execution of workflow and the accumulative stochastic distributions of tasks.

The rest of this article is organized as follows. Section 2 discusses the related work. The system model of workflow is introduced in Sect. 3. The definitions of formulas and symbols are explained in Sect. 4. Section 5 introduces Critical-Path-Performance-Evaluation VM selection strategy in detail. Experimental results and analysis are given in Sect. 6. Finally, Sect. 7 summarizes the paper.

2   Related Work

It is widely acknowledged that the problem of scheduling multi-task business processes with decentralized computing platforms is NP-hard [13,14]. Thus, heuristic and approximation-based methods can be good alternatives for generated sub-optimal results with affordable time-complexity.

For instance, Casas et al. [15] incorporated a bio-inspired strategy with the Efficient Tune-In (GA-ETI) mechanism for mapping workflow tasks into app. Wang et al. [16] leveraged a multi-stage game-theoretic framework. Its optimization scheme is yielded by multi-step dynamic game for generating the sub-optimal solutions with multi-constraints. These works can be limited due to the fact that they assumed time-invariant and static performance of infrastructural cloud services. They can thus be ineffective when dealing with cloud systems with unstable performance.

Various works assumed bounded performance instead of the static one. E.g., Mao et al. [17] introduced a VM-consolidation-based strategy, which is capable of finding near-optimal scheduling plans for consolidating distributed VMs and assumes bounded task response time. Calheiros et al. [18] assumed soft constraints and considered a critical path identification algorithm that exploits idle time-slots for high system utilization. They considered bounded response time of VMs as well. Poola et al. [19] developed a fault-tolerant and performance-change-aware workflow scheduling method, where performance variation is assumed to be bounded by a given distribution type. Ghosh et al. [20,21] assumed VM response time to be with an exponential distribution. While Zheng et al. [22] assumed a Pareto distribution as an approximation.

Imposing bounds effectively improves accuracy of models for cloud performance and lowers SLA violation rate. However, as explained in [22], doing so cloud bring pessimistic evaluation of system capabilities and causes low resource utilization.

Consider a cheap cloud service with time-varying QoS performance and with averaged/highest response time of 25 s/23 s and another expensive one with averaged/highest response time of 17 s/17.1 s. If 23 s is the timing constraint, the bound-based algorithm definitely selects the high-price one to avoid SLA violation. Nevertheless, 25 s occurs only in extreme cases and a smarter algorithm is supposed to decide the future tendency of breaching the 23 s threshold and choose the expensive VM only when such tendency is high.

Recently, various contributions considered time-variant or stochastic performance of cloud infrastructures and resources in their scheduling models. For example, Sahni *et al.* [23] proposed a performance-fluctuation-aware and cost-minimization method for dynamic multi-workflow scheduling with the constraint of completion time. It considers the fluctuation of performance and generates the scheduling scheme in real time. Li *et al.* [24] proposed a time-series-prediction-based approach to scheduling workflow tasks upon IaaS clouds. The fed the predicted performance trend into a genetic algorithm (GA) for yielding future scheduling plans. Haidri *et al.* [25] proposes a cost-effective deadline-time-aware (S-CEDA) resource scheduler that incorporates the expected value and variance of task processing time and inter-task communication time into workflow scheduling to optimize the total execution price and total execution time of the workflow. Our work differs from these in that we consider both real-time scheduling and cumulative stochastic distribution of performance. It estimates the cumulative stochastic distributions of workflow tasks in real time according to the critical path, and generates the scheduling scheme with the goal of reducing cost under the constraint condition.

3   System Model

A workflow $W = (T, E)$ is described by a Directed-Acyclic-Graph (DAG) where $T = \{t_1, t_2, \ldots, t_m\}$ denotes the set of tasks and E denotes the set of edges. The edge $e_{ij} \in E$ indicates the constraint between t_i and t_j $(i \neq j)$. This means that task t_j can only be started after t_i is completed. D identifies the pre-specified constraint.

VMs exist in the provider resource pool and are with different computing performance and price. If task t_i and t_j are run on different VMs, and $e_{ij} \in E$, a transmission of data between them is needed.

The charging plan is based on the pay-per-use billing pattern. Resource providers charge users based on the duration they spend for occupying VMs.

The execution order of a DAG is formally expressed by associating an index to every task. The index has a rage from 1 to m and the ith item indicates the order of running t_i. The above sequence configuration can be decided by the function $l : T \rightarrow N^+$ and encoded as a vector of a permutation of 1 to n. If i precedes k in the order, it doesn't necessarily mean that t_i precedes t_k unless they are on the same virtual machine. Workflow tasks performed by varying types of VMs often exhibit different performance. Moreover, the performance of the same virtual machine executing the same task fluctuates at different times as mentioned before. In order to meet the Qos constraints when the performance of virtual machines fluctuates, dynamic scheduling methods and performance fluctuation data should be considered.

4 Definition and Formulation

Symbols	Description
C	The cost of executing a workflow
D	The deadline constraint of a workflow
T	The set of tasks of a workflow
K	The set of VM types
τ	Workflow completion time
n	The number of tasks
$^*t_i$	The preceding sets of t_i
t_i^*	The subsequent sets of t_i
t_i	The ith task of a workflow
v_j	The jth VM instance
$h(j)$	A function to indicate cost-per-unit-time of VM v_j
$w(i)$	A function to indicate the VM to which task t_i is to be scheduled
$\overline{R_i^j}$	The average execution time of t_i-type task execute on v_j-type virtual machine
R_i^j	The acutual execution time of t_i-type task execute on v_j-type virtual machine
x_{ij}	The data transfer time between t_i and t_j
θ_j	The expected idle time of v_j, v_j is estimated to be available at this time
β_j	The actual idle time of v_j, all tasks to be performed on v_j are finished
ξ_i^j	The estimated starting time of task t_i if it is scheduled into VM v_j
χ_i^j	The actual starting time of task t_i if it is scheduled into VM v_j
λ_i^j	The estimated finishing time of task t_i if it is scheduled into VM v_j

Fig. 1. Notations and meanings.

Cost-effectiveness and performance objectives usually contradicts with each other. Our proposed strategy reconciles them and generates cost-effective schedules with as-low-as-possible cost while meeting the performance threshold at real time. The formal problem can therefore be described as follows,

$$\min C = \sum_{i=1}^{|n|} (R_i^{w(i)}) * h(w(i)) \tag{1}$$

$$subject\ to\ \tau \leq D$$

where C indicates the total cost, D the completion-time threshold, $h(j)$ the cost-per-time for occupying VM v_j. Notations and meanings are shown in Fig. 1.

In this work, our method uses the accumulated stochastic distributions of tasks to dynamically schedule cloud workflows. The performance of different tasks on the virtual machine is what we concern, it is important to keep track of how the task is being performed. The performance of virtual machines tends to fluctuate, and the performance of tasks is different. We perform the Gauss Legendre algorithm task on different configured virtual machines, the details of the task and the configuration of the virtual machine are explained in the case

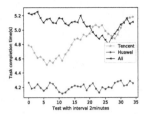
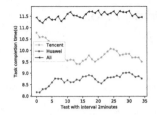
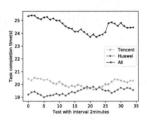

Fig. 2. Measured time for the Gauss Legendre algorithm to calculate 2 million digits of circumference ratio.

Fig. 3. Measured time for the Gauss Legendre algorithm to calculate 4 million digits of circumference ratio.

Fig. 4. Measured time for the Gauss Legendre algorithm to calculate 8 million digits of circumference ratio.

study section. Let R_i^j indicates the actual execution time of t_i-type task execute on v_j-type virtual machine and it's value fluctuates as shown in Fig. 2, 3 and 4.

F_{R_x} indicates the cumulative distribution function (CDF) of execution time. Let R_i^y, R_j^z denote the execution time of two tasks executed sequentially. The cumulative distribution function (CDF) of R_{i+j} is obtained by a convolution operation (which takes the CDFs of two random variables as input and generates the CDF of the sum of the two random variables, usually denoted by a symbol in math) of R_i^y and R_j^z:

$$
\begin{aligned}
F_{R_x}(t) &= Prob(R_x \leq t) \\
&= F_{R_i+R_j}(t) \\
&= F_{R_i} * F_{R_j}(t) \\
&= \int_0^\infty F_{R_i}(t - s) \times F_{R_j}(t)(s) \\
&= \int_0^\infty Prob\{R_i^y <= t - s\} \times Prob\{R_j^z <= s\}ds
\end{aligned}
\tag{2}
$$

We use the data in Fig. 2 as an example, and the result is shown in Fig. 5.

The estimated start time of the task is an important parameter that our method needs to use. ξ_i^j indicates the estimated beginning time of task t_i if it is allocated into VM v_j, calculated by the estimated completion time of the parent tasks, the transmission time, and the estimated idle duration. The virtual machine is idle when all tasks to be performed on VM are finished and the virtual machine is now available. ξ_i^j is used as the start time of the critical path from a task, which is dynamically updated to generate a more appropriate scheduling scheme during the task execution.

$$
\xi_i^m =
\begin{cases}
\max_{\substack{t_j \in {}^*t_i \\ n \neq m}} (\lambda_i^n + x_{j_i}, \lambda_j^m, \theta_m) & \text{if } {}^*t_i \neq \emptyset \\
\theta_m & \text{if } {}^*t_i = \emptyset
\end{cases}
\tag{3}
$$

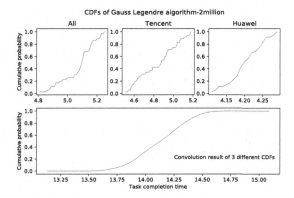

Fig. 5. A convolution sample of CDF

λ_i^j indicates the estimated ending time of task t_i if t_i is allocated into VM v_j. It is calculated by task's beginning time and the average performance of VMs. $\overline{R_i^j}$ indicates the average execution time, determined by the history log. If the type of task has not been performed, and there is no history, it is calculated based on the number of task instructions and virtual machine configuration. λ_i^j is used to get the estimated start time of the child task.

$$\lambda_i^j = \begin{cases} \chi_i^j + \overline{R_i^j} \text{ (if } t_i \text{ is in execution)} \\ \xi_i^j + \overline{R_i^j} \\ \text{(if } t_i \text{ is waiting for execution)} \end{cases} \tag{4}$$

As mentioned earlier, the execution time of the task and the performance of the virtual machine fluctuate. In order to dynamically adjust the allocation plan based on the current state, we need to update the expected start time of the task in the scheduling scheme through the actual execution of the task.

$$\chi_i^m = \begin{cases} \overset{t_j \in {}^* t_i}{\underset{n \neq m}{\max}} \{\chi_j^n + R_j^n + x_{j_i}, \chi_j^m + R_j^m, \beta_m\} \\ \qquad\qquad\qquad \text{if } {}^* t_i \neq \emptyset \\ \beta_m \qquad\qquad\qquad \text{if } {}^* t_i = \emptyset \end{cases} \tag{5}$$

χ_i^j indicates the actual beginning time of task t_i on condition that it is allocated into VM v_j. This means that v_j begins after it accepts all the dependent data of t_i. Note that the performance of the virtual machine fluctuates, the value of R is the actual execution time.

5   Critical-Path-Performance-Evaluation VM Selection Strategy

Algorithm 1: This pseudo-code illustrates the process of finding the critical path. This function starts with t_i looking for the path with the largest computation task, until it identifies the task with no child tasks. We considered the accumulated stochastic distributions of tasks of critical path tasks to calculate the duration of the entire workflow. When multiple termination tasks are existent, e.g., $t_j{}^* = \emptyset$ and $t_m{}^* = \emptyset$, this function is also applicable. To simplify the function, the variable $flagSize$ is used, which is not returned when the final result is generated.

Algorithm 1. CriticalPath

---

input: Task t_i
output: The critical path from t_i
1: $flagSize = 0$
2: $flagTask = null$
3: $flagPath = null$
4: **if** $t_j{}^* = \emptyset$ **then**
5:     $tempSize \Leftarrow taskSize(t_i)$
6:     $tempPath = Stack.push(t_i)$
7:        **return** $tempSize, tempPath$
8: **end if**
9: **if** $t_i{}^* \neq \emptyset$ **then**
10:     **for** $t_j \in t_i{}^*$ **do**
11:         $tempSize, tempPath = criticalPath(t_j)$
12:         $tempSize = taskSizeArr(t_j) + tempSize$
13:         **if** $tempSize > flagSize$ **then**
14:             $flagSize = tempSize$
15:             $flagTask = t_j$
16:             $flagPath = tempPath$
17:         **end if**
18:     **end for**
19:     $tempPath = Stack.push(flagTask)$
20:        **return** $flagSize, flagPath$
21: **end if**

Algorithm 2: This is the main part of the scheduling algorithm, we choose the virtual machine based on the accumulated stochastic performance distributions of critical path. The critical path starting time from each task is different, which can be obtained from (3). If there are no virtual machines that meet the constraints, reduce the constraints or simply use the better and more expensive virtual machines. Parameter $taskQue$ is the queue of tasks that can be allocated in parallel. Parameter vm_pool is the set of VMs that we can used in scheduling algorithm.

Algorithm 3: The accumulated stochastic performance distributions of critical path is analyzed and calculated. This function uses the formula (2) approach to consider the performance of the critical path. F_{R_x} is the cumulative distribution function (CDF) of the performance of the x-type task or set of path tasks in the specified virtual machine. For tasks that do not record performance, it is calculated from the configuration of the virtual machine and the details of the task.

Algorithm 4: In the process of making allocation decision, the actual execution duration of tasks can differs from that of expected. As the task runs, the performance estimation should thus be updated. When a task starts executing, its expected completion time is updated by the specific start time of the task. According to the actual execution of the task, a new schedulable list is generated to calculate the new and more accurate scheduling scheme.

Algorithm 2. Scheduling

---

input: Task queue: $taskQue$, VM pool: vm_pool
output: Scheduling scheme
1: $flagVM = null$
2: $flagCost = INF$
3: **while** $taskQue \neq \emptyset$ **do**
4:     $t_i = Queue.deque(taskQue)$
5:     $Cpath = CriticalPath(t_i)$
6:     $CDFs = VM\_PathPerfrom(Cpath)$
7:     **for** $v_j \in vm\_pool$ **do**
8:         $F_{R_x} = CDFs(v_j)$
9:         //Gets the critical path performance of v_j
10:         $tempTime = D - \xi_i^j$
11:         $Prob = F_{R_x}(tempTime)$
12:         //$limit$ is the degree of constraint
13:         **if** $Prob \geq limit$ **then**
14:             $tempCost = F_{R_i^j}(Prob) * h(j)$
15:             **if** $flagCost < tempCost$ **then**
16:                 $flagCost = tempCost$
17:                 $flagVM = v_j$
18:             **end if**
19:         **end if**
20:     **end for**
21:     **if** $flagVM == null$ **then**
22:         $flagVM = max(h(x))$
23:     **end if**
24:     $schedules.add(t_i, v_j)$
25:     $Update(taskQue)$
26: **end while**

Algorithm 3. VM_PathPerform

---

input: Critical Path: *path*
output: CDFs of Critical Path: *CDFs*
1: **for** $k \in K$ **do**
2:      $tempPath = path$
3:      $t_i = Stack.pop(tempPath)$
4:      $F_{R_x} \Leftarrow Perform(t_i, k)$
5:      // If $F_{R_x} == null$ do $F_{R_x} = Prob(\overline{R_i^k} = 1)$
6:      **for** $t_j = Stack.pop(tempPath)$ **do**
7:          $F_{R_j} \Leftarrow Perform(t_j, k)$
8:          $F_{R_x} = F_{R_x} * F_{R_j}$
9:      **end for**
10:     $CDFs.append(F_{R_x})$
11: **end for**
12: **return** $CDFs$

Algorithm 4. Excution

---

input: Workflow: T
output: Update related parameters
1: **while** T not finished **do**
2:      $t_i, v_j = schedules.get()$//Scheduling scheme
3:      Wait $^*t_i$ all finished
4:      Start executing t_i
5:      // When t_i start running, update relevant data
6:      $Update(\chi_i^j)$
7:      $Update(\lambda_i^j)$
8:      $Update(taskQue)$
9:      End of execution of task t_i
10:     // When t_i is completed, update data $F_{R_i^j}$
11:     $Update(F_{R_i^j})$
12: **end while**

---

6   Case Study

In this part, we perform a case study of a real-world science workflow deployed on a commercial IaaS cloud to compare the traditional scheduling approach to our proposed framework.

We consider different classical scientific workflow templates, namely **Montage**, **CyberShake**, and **Epigenomics** as shown in Fig. 6 to support tasks of Gauss Legendre calculations with a different number of digits. The GaussLegendre calculation is a highly-memory-requiring iterative procedure to compute the digits of circumference ratio to a specific number of digits. This procedure is implemented by a benchmark tool, i.e., Super-Pi (from http://www.superpi. net/). This tool is frequently used in testing floating-point performance of computing systems. Tasks of the sample workflows are required to run the Super-Pi tests with different requirements of the numbers of digits to generate as given in Table 1.

Table 1. Tasks of three scientific workflows(million).

Montage	t_1	t_2	t_3	t_4	t_5	t_6	t_7	t_8	t_9	t_{10}
digits of circumference ratio	2	4	8	2	4	8	2	4	8	2
Montage	t_{11}	t_{12}	t_{13}	t_{14}	t_{15}	t_{16}	t_{17}	t_{18}	t_{19}	t_{20}
digits of circumference ratio	4	8	2	4	8	2	4	8	2	4
Montage	t_{21}	t_{22}	t_{23}	t_{24}						
digits of circumference ratio	8	2	4	8						
CyberShake	t_1	t_2	t_3	t_4	t_5	t_6	t_7	t_8	t_9	t_{10}
digits of circumference ratio	2	4	8	2	4	8	2	4	8	2
CyberShake	t_{11}	t_{12}	t_{13}	t_{14}	t_{15}	t_{16}	t_{17}	t_{18}	t_{19}	t_{20}
digits of circumference ratio	4	8	2	4	8	2	4	8	2	4
Epigenomics	t_1	t_2	t_3	t_4	t_5	t_6	t_7	t_8	t_9	t_{10}
digits of circumference ratio	2	4	8	2	4	8	2	4	8	2
Epigenomics	t_{11}	t_{12}	t_{13}	t_{14}	t_{15}	t_{16}	t_{17}	t_{18}	t_{19}	t_{20}
digits of circumference ratio	4	8	2	4	8	2	4	8	2	4

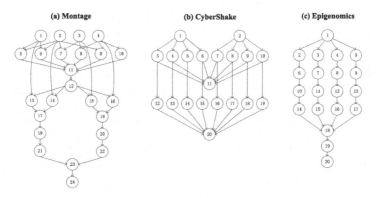

Fig. 6. Overview of there workflow templates for the case study.

We employ three commercial cloud services, namely Ali, Huawei, and Tencent to provide supporting VMs for the workflow tasks. VMs from Ali, Huawei, and Tencent clouds are with different resource configurations, i.e., 1 g RAM /2.5 GHz(1-core)/40 G storage for Ali, 2 g RAM/2.6 GHz(1-core)/50 G storage for Huawei, and 2 g RAM/2.5 GHz(1-core)/50 G storage for Tencent. The cost-per-second of these services are 1.20, 1.60, and 1.40 cents, according to their charging plans, respectively. The maximum number of required VMs equals that of tasks and VMs are available from the beginning to the end. We tested these VMs by using Sugon I450 server (4-CPU Intel Xeon 5506/128 G RAM) at 2 2 min fixed intervals. The task completion time for different virtual machines to calculate circumference ratio reaching a varying number of decimal digits shown as Fig. 2, 3 and 4. We also used other Sugon I450 server to generate the scheduling scheme.

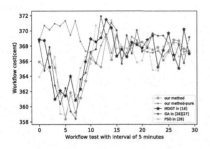

Fig. 7. Comparison of cost of Montage workflow.

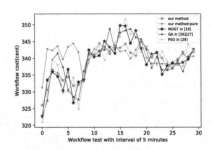

Fig. 8. Comparison of cost of Cyber-Shake workflow.

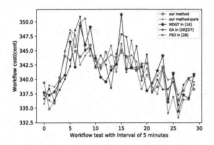

Fig. 9. Comparison of cost of Epigenomics workflow.

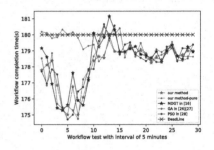

Fig. 10. Comparison of completion time of Montage workflow.

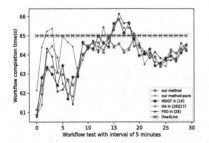

Fig. 11. Comparison of completion time of CyberShake workflow.

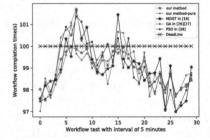

Fig. 12. Comparison of completion time of Epigenomics workflow.

We calculated the workflow execution completion time roughly according to the maximum-task-path from DAGs and previously measured data. The baseline execution time for Montage, Cybershake, and Epigenomic workflows should fall into [136;226], [56;72], and [88;112] secedes. We also used the previously measured data as the initial performance record for our proposed method, and the timing thresholds of Montage, CyberShake, and Epigenomics are 180 s, 65 s, and 100 s. To compare, we employ GA [26,27], PSO [28] and MDGT [16] as the baseline algorithms. To validate the effectiveness of the stochastic-performance-distribution-based framework, we consider a pure version of our proposed method without pre-measured data, where CDFs is collected dynamically at execution time.

The scheduling scheme executes the workflow at five-minute intervals, which is a safe time to ensure that no other Gauss Legendre tasks run on the virtual machine. We set the constraint parameter *limit* in our method to 0.95, requiring that the critical path execution time starting from each task has a probability of at least 95% meeting Qos conditions. If no virtual machine exists that satisfies the constraint, we use the fastest but more expensive virtual machine to reduce the workflow completion time.

As can be observed from Figs. 7, 8 and 9, our method beats PSO, GA, and MDGT in terms of average cost (as shown in Figs. 13, the averaged cost of our method for Montage/Cybershake/Epigenomic are 365.7/338.5/341.2 cents, while those of GA are 366.8/339.2/341.79 cents, those of PSO for are 366.2/338.7/341.76 cents, those of MDGT are 366.5/338.9/341.7 cents).

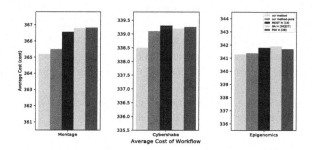

Fig. 13. The average cost of multiple executions of workflow.

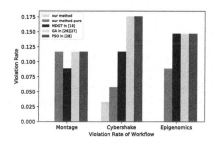

Fig. 14. Violation rate of workflow.

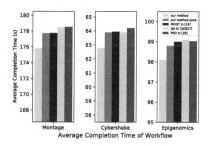

Fig. 15. Average completion time.

As can be seen from Figs. 14–15 as well, our method clearly achieves lower average completion times by 1.1%, 2%, and 0.9% for three workflows, respectively. It's worth noting that such reduced completion times also lead to fewer violations to the deadline constraints. (the violation rates of our method for Montage/Cybershake/Epigenomic are 0%/3%/0%, while those of GA are 11.7%/17.6%/14.7%, those of PSO are 11.7%/17.6%/14.7%, those of MDGT are 8.9%/11.7%/14.7%). Our method beats the pure one as well (the violation rates of our method for Montage/Cybershake/Epigenomic are 0%/0%/0%, while those of the our method-pure one are 11.7%/5.8%/8.9%). By comparing the results of our method and our method-pure as shown in Fig. 7, 8, 9, 10, 11 and 12, the results show that they become approximately consistent after an indeterminate time that executing 13/10/9 times for Montage/Cybershake/Epigenomic workflow. The proposed method can obtain sufficient and effective performance evaluation data in a relatively short time.

7   Conclusion

In this work, we develop a novel Stochastic-Performance-Distribution-Based approach for scheduling workflows upon clouds infrastructures. The developed strategy is capable of modeling time-varying performance of cloud resources and generating cost-effective scheduling plans to reduce monetary cost while following constraints of Service-Level-Agreement (SLA). It is featured by a stochastic-performance-distribution-based framework for estimation and optimization of workflow critical paths in terms of the accumulated stochastic distributions of tasks. Comprehensive tests based on commercial clouds and multiple well-known scientific workflow templates show that our method beats other traditional ones.

Acknowledgement. This work is supported in part by Science and Technology Program of Sichuan Province under Grant 2020JDRC0067.

References

1. Xia, Y., Zhou, M., Luo, X., Pang, S., Zhu, Q.: Stochastic modeling and performance analysis of migration-enabled and error-prone clouds. IEEE Trans. Ind. Inform. **11**(2), 495–504 (2015)
2. He, Q., et al.: Spectrum-based runtime anomaly localisation in service-based systems. In: 2015 IEEE International Conference on Services Computing, SCC 2015, New York City, NY, USA, 27 June–2 July 2015, pp. 90–97 (2015)
3. Gai, K., Qiu, M., Zhao, H., Tao, L., Zong, Z.: Dynamic energy-aware cloudlet-based mobile cloud computing model for green computing. J. Netw. Comput. Appl. **59**, 46–54 (2016)
4. Wang, Y., He, Q., Ye, D., Yang, Y.: Formulating criticality-based cost-effective fault tolerance strategies for multi-tenant service-based systems. IEEE Trans. Softw. Eng. **44**(3), 291–307 (2018)
5. Gai, K., Qiu, M., Zhao, H.: Energy-aware task assignment for mobile cyber-enabled applications in heterogeneous cloud computing. J. Parallel Distrib. Comput. **111**, 126–135 (2018)

6. Wu, Q., Ishikawa, F., Zhu, Q., Xia, Y., Wen, J.: Deadline-constrained cost optimization approaches for workflow scheduling in clouds. IEEE Trans. Parallel Distrib. Syst. **28**(12), 3401–3412 (2017)

7. Haddad, J.E., Manouvrier, M., Rukoz, M.: TQoS: Transactional and QoS-aware selection algorithm for automatic web service composition. IEEE Trans. Serv. Comput. **3**(1), 73–85 (2010)

8. Wang, Y., et al.: Multi-objective workflow scheduling with deep-q-network-based multi-agent reinforcement learning. IEEE Access **7**, 39974–39982 (2019)

9. Gai, K., Choo, K.R., Qiu, M., Zhu, L.: Privacy-preserving content-oriented wireless communication in internet-of-things. IEEE Internet Things J. **5**(4), 3059–3067 (2018)

10. Schad, J., Dittrich, J., Quiané-Ruiz, J.: Runtime measurements in the cloud: observing, analyzing, and reducing variance. PVLDB **3**(1), 460–471 (2010)

11. Jackson, K.R., Ramakrishnan, L., Muriki, K., Canon, S., Wright, N.J.: Performance analysis of high performance computing applications on the amazon web services cloud. In: Proceedings of the Cloud Computing, Second International Conference, CloudCom 2010, Indianapolis, Indiana, USA, 30 November–3 December 2010 (2010)

12. Chen, H., Zhu, X., Liu, G., Pedrycz, W.: Uncertainty-aware online scheduling for real-time workflows in cloud service environment. IEEE Trans. Serv. Comput. (2018). https://doi.org/10.1109/TSC.2018.2866421

13. Ibarra, O.H., Kim, C.E.: Heuristic algorithms for scheduling independent tasks on nonidentical processors. J. ACM **24**(2), 280–289 (1977)

14. He, Q., et al.: QoS-aware service selection for customisable multi-tenant service-based systems: Maturity and approaches. In: 8th IEEE International Conference on Cloud Computing, CLOUD 2015, New York City, NY, USA, 27 June–2 July 2015, pp. 237–244 (2015)

15. Casas, I., Taheri, J., Ranjan, R., Wang, L., Zomaya, A.Y.: GA-ETI: an enhanced genetic algorithm for the scheduling of scientific workflows in cloud environments. J. Comput. Sci. **26**, 318–331 (2018)

16. Wang, Y., Jiang, J., Xia, Y., Wu, Q., Luo, X., Zhu, Q.: A multi-stage dynamic game-theoretic approach for multi-workflow scheduling on heterogeneous virtual machines from multiple infrastructure-as-a-service clouds. In: Ferreira, J.E., Spanoudakis, G., Ma, Y., Zhang, L.-J. (eds.) SCC 2018. LNCS, vol. 10969, pp. 137–152. Springer, Cham (2018). https://doi.org/10.1007/978-3-319-94376-3_9

17. Mao, M., Humphrey, M.: Auto-scaling to minimize cost and meet application deadlines in cloud workflows. In: Conference on High Performance Computing Networking, Storage and Analysis, SC 2011, Seattle, WA, USA, 12–18 November 2011, pp. 49:1–49:12 (2011)

18. Calheiros, R.N., Buyya, R.: Meeting deadlines of scientific workflows in public clouds with tasks replication. IEEE Trans. Parallel Distrib. Syst. **25**(7), 1787–1796 (2014)

19. Poola, D., Garg, S.K., Buyya, R., Yang, Y., Ramamohanarao, K.: Robust scheduling of scientific workflows with deadline and budget constraints in clouds. In: 28th IEEE International Conference on Advanced Information Networking and Applications, AINA 2014, Victoria, BC, Canada, 13–16 May 2014, pp. 858–865 (2014)

20. Ghosh, R., Longo, F., Frattini, F., Russo, S., Trivedi, K.S.: Scalable analytics for IaaS cloud availability. IEEE Trans. Cloud Comput. **2**(1), 57–70 (2014)

21. Yin, X., Ma, X., Trivedi, K.S.: An interacting stochastic models approach for the performance evaluation of DSRC vehicular safety communication. IEEE Trans. Comput. **62**(5), 873–885 (2013)

22. Zheng, W., et al.: Percentile performance estimation of unreliable IaaS clouds and their cost-optimal capacity decision. IEEE Access **5**, 2808–2818 (2017)

23. Sahni, J., Vidyarthi, D.P.: A cost-effective deadline-constrained dynamic scheduling algorithm for scientific workflows in a cloud environment. IEEE Trans. Cloud Comput. **6**(1), 2–18 (2018)

24. Li, W., Xia, Y., Zhou, M., Sun, X., Zhu, Q.: Fluctuation-aware and predictive workflow scheduling in cost-effective infrastructure-as-a-service clouds. IEEE Access **6**, 61488–61502 (2018)

25. Haidri, R.A., Katti, C.P., Saxena, P.C.: Cost-effective deadline-aware stochastic scheduling strategy for workflow applications on virtual machines in cloud computing. Concurr. Comput. Pract. Exp. **31**(7), e5006.1–e5006.24 (2019)

26. Meena, J., Kumar, M., Vardhan, M.: Cost effective genetic algorithm for workflow scheduling in cloud under deadline constraint. IEEE Access **4**, 5065–5082 (2016)

27. Zhu, Z., Zhang, G., Li, M., Liu, X.: Evolutionary multi-objective workflow scheduling in cloud. IEEE Trans. Parallel Distrib. Syst. **27**(5), 1344–1357 (2016)

28. Rodriguez, M.A., Buyya, R.: Deadline based resource provisioning and scheduling algorithm for scientific workflows on clouds. IEEE Trans. Cloud Comput. **2**(2), 222–235 (2014)

An FM Developer Recommendation Algorithm by Considering Explicit Information and ID Information

Xu Yu[1], Yadong He[1], Biao Xu[2], Junwei Du[1(✉)], Feng Jiang[1(✉)], and Dunwei Gong[3]

[1] School of Information Science and Technology,
Qingdao University of Science and Technology, Qingdao 266061, China
djwqd@163.com, jiangkong@163.net
[2] Department of Electronic Engineering, Shantou University, Shantou 515063, China
[3] School of Information and Control Engineering, China University of Mining and Technology,
Xuzhou 221116, Jiangsu, China

Abstract. Recently, the developer recommendation on crowdsourcing software platform is of great research significance since an increasingly large number of tasks and developers have gathered on the platforms. In order to solve the problem of cold-start, the existing developer recommendation algorithms usually only use explicit information but not ID information to represent tasks and developers, which causes poor performance. In view of the shortcomings of the existing developer recommendation algorithms, this paper proposes an FM recommendation algorithm based on explicit to implicit feature mapping relationship modeling. This algorithm firstly integrates fully the ID information, explicit information and rating interaction between the completed task and the existing developers by using FM algorithm in order to get the implicit features related to their ID information. Secondly, for the completed tasks and existing developers, a deep regression model is established to learn the mapping relationship from explicit features to implicit features. Then, for the cold-start task or the cold-start developer, the implicit features are determined by the explicit features according to the deep regression model. Finally, the ratings in the cold-start scene can be predicted by the trained FM model with the explicit and implicit features. The simulation results on Topcoder platform show that the proposed algorithm has obvious advantages over the comparison algorithm in precision and recall.

Keywords: Crowdsourcing software development · Developer recommendation · Cold-start problem · Deep regression model · FM algorithm

1 Introduction

In recent years, more and more software companies have begun to adopt crowdsourcing software engineering model [1] for software development, which is the application of crowdsourcing concept in the field of software development. With the rapid development of various crowdsourcing software platforms, the number of publishing tasks and registered developers has increased dramatically. Therefore, the problem of "information

© Springer Nature Switzerland AG 2020
W.-S. Ku et al. (Eds.): ICWS 2020, LNCS 12406, pp. 49–60, 2020.
https://doi.org/10.1007/978-3-030-59618-7_4

overload" on crowdsourcing software platforms is becoming more and more serious, which makes the tasks and developers face a serious problem in selection. In this context, developer recommendation is of paramount importance in research and application, and has attracted the attention of some researchers in recent years.

Mao et al. [2] first extract task features, and then use classification algorithm to match the task features with the ID number of the winning developer. Shao et al. [3] firstly train the neural network model based on the category and numerical attributes of the task features, then train the Latent Semantic Index (LSI) model by using the task description attribute, and finally complete the developers' recommendations by combining the two models. These methods only use task explicit information for input feature presentation, and do not take into account other rich heterogeneous information on crowdsourcing software development platform. Hence, the recommendation performance is not ideal. Besides, they cannot solve the problem of developers' cold-start since they use developer ID as the class label.

Zhu et al. [4] regard the problem of developer recommendation as one of information retrieval. This method is similar to a classification model in essence, which takes the explicit features of tasks and developers as model input, and the evaluation results ("recommended" or "not recommended") as label information. Because only the explicit features of tasks and developers are used as the model input, the method can solve the cold-start problem. However, similar to the previous methods, this method still fails to fully represent explicit information on the crowdsourcing software development platform. More importantly, in order to solve the cold-start problem, this method only takes the explicit features as input, and does not take into consideration the task and developer ID information, so the effect is not ideal. In fact, according to the Factorization Machines algorithm (FM) [5], the ID information can reflect the implicit features that are closely related to the rating results. Hence it is an important kind of feature that needs to be considered in the recommendation system. Although FM model can integrate high-dimensional sparse explicit features and ID features, the model cannot solve the problem of cold-start. Therefore, it cannot be directly applied to developer recommendation scenarios.

Aiming at the above shortcomings, this paper proposes an FM recommendation algorithm based on explicit to implicit feature mapping relationship modeling for the cold-start scenario on crowdsourcing software platform. On one hand, this method extracts and represents the features related to tasks and developers according to the explicit description information; constructs a fine-grained rating matrix according to the scoring information. Besides, the ID information is represented with one-hot encoding. On the other hand, FM algorithm is firstly used to fuse the explicit features, ID information and corresponding ratings. Secondly, based on the trained FM model, the implicit features of existing tasks and developers are acquired, and the mapping relationship from explicit features to implicit features is established based on the deep regression model. Then, based on the mapping relationship, the implicit features are calculated according to the explicit features of cold-start tasks or developers. Finally, according to the explicit and implicit features of cold-start tasks and developers, the trained FM model is used for rating prediction. Extensive experiments on the data set from Topcoder show that it has obvious advantages in precision and recall compared with the comparison algorithm, which shows the effectiveness of this method.

The rest of this paper is organized as follows: Sect. 2 reviews the related works. In Sect. 3, an FM recommendation algorithm based on explicit to implicit feature mapping relationship modeling is proposed. In Sect. 4, extensive comparative experiments are carried out, and the experimental results are analyzed in detail. Section 5 summarizes the whole paper and proposes future research orientations.

2   Related Work

2.1   Traditional Recommendation

In order to solve the problem of information overload in the context of big data and meet the personalized needs of different users, the recommendation algorithm has been widely used. Two of the most famous recommendation algorithms are content-based (CB) [6] and collaborative filtering (CF) [7]. According to the attribute information of the user's favorite items in the past, CB algorithm recommends the similar items for the user. However, CB algorithm usually needs to represent the features of items. Inaccurate or insufficient feature representation will seriously affect its recommendation performance. By contrast, CF algorithm does not rely on the content of goods, only uses the feedback of users to items to mine the preferences of users. The recommendation results have better diversity, but the problems of cold-start and data sparsity are the bottlenecks of CF algorithm. In addition, hybrid recommendation [8] and recommendation algorithm [9] based on deep learning technology have been widely studied in recent years, which provides new research ideas for the field of recommendation algorithm.

2.2   Developer Recommendation in Crowdsourcing Software Engineering

Compared with the traditional recommendation system, the solution of cold-start problem is particularly important for developer recommendation. Therefore, simple applications of traditional recommendation algorithm cannot effectively solve the developer recommendation problem. In the literature of developer recommendation, the first systematic study comes from Mao et al. [2]. Following, Shao et al. [3] propose a NNLSI model. However, as both of these two models fail to fully represent the input features, the recommendation performance is not ideal. Besides, both of the two models cannot solve the developer cold-start problem.

Zhu et al. [4] propose a recommendation model based on Learning to Rank (LR). The method only designs a small number of features based on the task and developer description information, which affects the recommendation performance of the model. In addition, it only considers the explicit features as the input information, but fails to use the implicit features reflected by the ID information. Therefore, the recommendation effect is still not ideal.

3 The Proposed Model

3.1 Feature Representation

Task Explicit Feature Representation. Software crowdsourcing tasks contain various types of features, ranging from text features to numeric features. The main text features include "Title", "Task Description", "PL" (Programming Language), and "Tech" (Techniques). The main numeric features are "Date" (task post date), "Duration" (allocated task duration), and "Payment" [2]. Next, we will give the representation and calculation methods of the features.

(1) Date, Duration, and Payment
    The above three numeric features can be obtained from the platform directly. In order to eliminate the influence of different scales, Z-score normalization is used to process the numeric feature values.
(2) PL and Tech
    Since the value of PL and Tech are always composed of several labels provided in the platform, we can naturally use a binary vector with a dimension equal to the number of labels to represent it, where value 1 denotes the developer has the corresponding skill and 0 otherwise.
(3) Title and Task description
    In this paper, we utilize Bert [10] model to encode the title and task description. Moreover, we use PCA to reduce dimension.

Developer Explicit Feature Representation. As to the features of developers, we consider static and dynamic features. The feature details of developers are presented in Table 1. Next, we will give the representation and calculation methods of the features.

(1) Skill
    Like PL and Tech features of the tasks, we also utilize a binary vector to represent it.
(2) Self-description
    Self-description information is represented in the same way as the 'Title' and 'Task description' features of the tasks.
(3) Developer type and Location
    We use the one-hot encoding to represent these two categorical features.
(4) The numeric features
    The numeric features can be obtained from the platform directly, which are preprocessed in the same way as the numeric features of tasks.

Task and Developer ID Information Representation. We use one-hot encoding to represent task and developer ID information.

Table 1. The feature details of developers

Type	Feature	Format	Description
Static	Skill	Text	Skills that developers are good at
	Developer type	Categorical	DATA_SCIENCE, DEVELOP, DESIGN
	Location	Categorical	Developer's work city
	Registration date	Numeric	The registration date of developers
	Self-description	Text	Self-description of developers
Dynamic	Number of tasks completed	Numeric	The number of tasks completed
	Number of tasks won	Numeric	The number of tasks won
	Submission rate	Numeric	Task submission rate
	Total evaluation score	Numeric	Total evaluation score of platform experts on developers' historical completion tasks
	Activity	Numeric	Number of tasks completed in the past three months
	Reliability	Numeric	Platform measurement of developer credit
	Forum posts	Numeric	Developer's performance on the Q&A forum

Rating Matrix Construction. This paper takes Topcoder platform as an example to design the developer recommendation algorithm. For Topcoder platform, when the developers submit the results, the platform will organize experts to evaluate the results and grade all the submitted results, with the score range of [0,100]. In order to be consistent with the traditional recommendation scenarios, this paper uses the min-max normalization to normalize the percentage rating to the [1–5] range, and then constructs the task-developer-rating matrix R.

3.2 Training Based on FM Fusion Model

We regard both the explicit features and ID features of tasks and developers as the input features of regression problem and the rating as the regression variable. In the model training stage, FM model is used to fuse the explicit features, ID features and ratings. This paper uses second order FM model to fuse heterogeneous information. For developer recommendation scenarios, see Eq. (1) for its expression.

$$\hat{r} = w_0 + \sum_{i=1}^{n_1+m_1+n+m} w_i x_i + \sum_{i=1}^{n_1+m_1+n+m} \sum_{j=i+1}^{n_1+m_1+n+m} w_{ij} x_i x_j \tag{1}$$

Among them, model parameters w_0, w_j and w_{ij} represent global bias, weight corresponding to feature i and weight of interaction term between feature i and feature j,

m and n represent the number of existing tasks and developers respectively, m_1 and n_1 represent the explicit feature dimensionality of tasks and developers respectively.

In Eq. (1), the weight w_{ij} of interaction terms between x_i and x_j can be expressed as $w_{ij} = v_i^T v_j$, where v_i and v_j represent the latent factor vector corresponding to feature x_i and x_j, respectively. The dimensionality k of the latent factor vector is usually specified manually by the user.

In order to train FM regression model, we minimize the regularized squared error.

$$(w_0^*, w^*, v^*) = \underset{w_0, w, v}{argmin} \left(\sum_{(x,y) \in D} (\hat{y}(x|w_0, w, v) - y)^2 + \lambda_{w_0} w_0^2 + \lambda_w \sum_{i=1}^{n_1+m_1+n+m} w_i^2 + \lambda_v \sum_{i=1}^{n_1+m_1+n+m} \sum_{f=1}^{k} v_{i,f}^2 \right) \quad (2)$$

where D denotes the training set, λ_{w_0}, λ_w, and λ_v are the regularization coefficients of three kinds of parameters to avoid overfitting.

It should be noted that FM model cannot deal with the cold-start problem caused by the ID feature. This can be explained from the optimization problem (2), as we cannot obtain the values of latent factor vectors corresponding to the cold-start tasks or cold-start developers. To facilitate the description, we first give the following definitions.

Definition 1. Latent factor vectors of tasks and developers. The latent factor vectors of tasks and developers are defined as the latent factor vectors in FM model corresponding to the feature whose value is 1 in the one-hot encoding of their ID numbers.

Definition 2. . Implicit features of tasks and developers. The implicit features of tasks and developers are defined as their latent factor vectors of tasks and developers.

As the cold-start tasks or developers do not have any score, their corresponding implicit features cannot be trained. Therefore, we cannot use Eq. (1) to predict the rating of the cold-start tasks to the developers. If there is a method to calculate the implicit features of cold-start tasks or cold-start developers, we can easily use the trained FM model to predict the score. We will give the detailed approach below.

3.3 The Cold-Start Problem

In the developer recommendation scenario, two situations need to be solved to deal with the cold-start problem: predicting the ratings of the cold-start tasks to the existing developers, and predicting the ratings of the cold-start tasks to the cold-start developers. We focus on the first case, and the second case is similar to the first one.

Predicting the Ratings of the Cold-Start Tasks to the Existing Developers. As the traditional FM model cannot model the implicit feature (v_i) of cold-start tasks during the training process, it cannot solve the problem of cold-start. Therefore, obtaining the implicit feature of cold-start tasks is the key to solve the problem. This paper attempts to use explicit features to induce implicit features. Next, we will model the mapping relation from explicit features to implicit features.

In this paper, we model the mapping relationship based on the deep regression network. First, we use the Stacked Denoising AutoEncoder (SDAE) [11] to transform the original explicit features of tasks to low-dimensional high-level features. Second, the linear regression model is constructed by using the low-dimensional high-level features as the input and the implicit features of tasks in the training set as the output.

(1) Dimension reduction

In this paper, SDAE is utilized to learn a good low-dimensional feature representation [11]. We set $k_1 > k_2 > k_3 > k_4$ to obtain high-level feature representation, shown in Fig. 1.

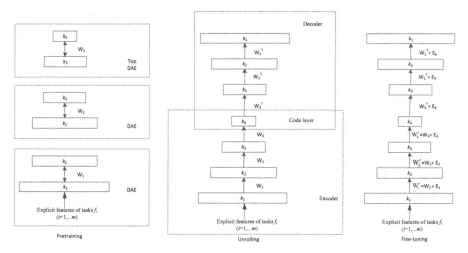

Fig. 1. The training of SDAE on explicit features of existing tasks.

(2) Linear Regression Analysis

A layer of traditional linear regression model is added to the outer layer of the stacked denoising autoencoder network, by which the low-dimensional high-level explicit features are transformed into implicit features. The linear regression unit does not contain any activation function, and only calculates the weighted sum of input units. The loss function can be defined as:

$$\frac{1}{m}\sum_{i=1}^{m}\left\|\hat{v}_i - v_i\right\|^2 \tag{3}$$

where v_i is the implicit feature value of task i, $\hat{v}_i$ is the implicit feature value of task i predicted by the deep regression model, and m is the number of existing tasks.

In this paper, during the training process of the deep regression model, we use the pre-trained weights to initialize the stacked denoising autoencoder, and use the random weights to initialize the traditional regression model in the outermost layer. Then, the back propagation algorithm is used to fine tune all weights, so as to obtain the final deep regression model.

(3) Predicting the ratings for existing developers on cold-start tasks

Suppose that the explicit features of the cold-start task i is f_i, and the mapping relation from the explicit features to the implicit features is F_1. According to the mapping relation, the implicit features v_i of the cold-start task i can be expressed as $v_i = F_1(f_i)$. Let f_j and v_j respectively denote the explicit features and implicit features of any existing developer j. According to the trained FM model, the ratings for developer j on cold-start task i can be calculated by substituting f_i, v_i, f_j, and v_j into the model.

Predicting the Ratings of the Cold-Start Tasks to the Cold-Start Developers. According to the proposed deep regression model, we can also obtain the mapping relation F_2, which transforms the explicit features of developers to implicit features. In the same way, the ratings for cold-start developer j on cold-start task i can be obtained.

4 Experiment

4.1 Introduction to the Topcoder Data Acquisition

Most of the tasks of Topcoder platform are published in the form of competitions. In this experiment, we crawl the contest task data related to software development, including system design, coding, module testing, etc. We collected historical tasks that have been crowdsourced between Oct. 2012 and Mar. 2019, from which we further filter out the incomplete data. In all, 520 tasks and 8286 developers are chosen as the original dataset.

4.2 Comparison Algorithms

At present, a few developer recommendation algorithms have been proposed, such as the model propose in [2], NNLSI [3] and LR [4]. However, the model proposed in [2] and NNLSI cannot solve the developer cold-start problem. LR is designed for Zhubajie website and cannot be applied to Topcoder, so we will not compare our method with the above three models.

In order to compare with the proposed MRFMRec algorithm, we consider a classical recommendation algorithm, FM algorithm, and apply it to the domain of developer recommendation. To enable it to deal with the problems of task cold-start and developer cold-start, the FM algorithm used for comparison only models the relation between task and developer's explicit features and ratings. As MRFMRec algorithm considers both the explicit features and ID features of tasks and developers, and the comparison method only considers the explicit features, the comparison results will fully reveal whether ID features are important for recommendation performance.

MRFMRec: FM algorithm is implemented by libFM software [5]. In FM algorithm, the dimensionality f of the latent factor vector is tried in $\{5, 10, 15, 20, 25, 30, 35, 40\}$, and the three regularization parameters λ_{w_0}, λ_w, λ_V are all tried in $\{0.001, 0.01, 0.1, 1, 10, 100\}$. We determine the optimal parameter value based on the test effect of FM algorithm on the validation set. For the deep regression model, we set $k_2 = \lfloor k_1/2 \rfloor$, $k_3 = \lfloor k_2/2 \rfloor$, $k_4 = \lfloor k_3/2 \rfloor$, where k_1 denotes the dimensionality of tasks or developers.

FM: The setting of relevant parameters in the algorithm is consistent with that in MRFMRec.

4.3 Metrics

We first use Mean Absolute Error (MAE) in Eq. (4) to evaluate the rating prediction precision of MRFMRec.

$$MAE = \frac{1}{|T_1|} \sum_{(t,d)\in T_1} \left| r_{td} - \hat{r}_{td} \right| \tag{4}$$

where r_{td} and $\hat{r}_{td}$ represent the ground truth rating and the predicted rating of task t to developer d, respectively, and T_1 is the test sample set.

Then, we use *Precision* and *Recall* to compare our MRFMRec model with FM. Let $R(t)$ be the predicted 'recommended' list obtained by a recommendation algorithm on a *tested developer list DL* = $\{d_1, d_2, ..., d_l\}$. Let $T(t)$ be the true 'recommended' list, i.e., the ground truth list on *DL*. Then, *Precision* and *Recall* are defined as:

$$Precision = \frac{\sum\limits_{t\in T_2} |R(t) \cap T(t)|}{\sum\limits_{t\in T_2} |R(t)|} \tag{5}$$

$$Recall = \frac{\sum\limits_{t\in T_2} |R(t) \cap T(t)|}{\sum\limits_{t\in T_2} |T(t)|} \tag{6}$$

where T_2 is the test sample set.

Note that the *tested developer list DL* can be represented by $T_E(t)$ in the following

$$DL = T(t) \cup \bar{T}(t) \tag{7}$$

where $\bar{T}(t) = DL - T(t)$ denotes the true 'not recommended' developer from the *tested developer list*. Hence, *DL* should contain both 'recommended' and 'not recommended' developers. How to design *DL* will be given in the following.

4.4 Data Preparation

We choose the rating matrix and relevant information composed of 520 tasks and 8286 developers as the experimental data. In order to evaluate the recommended performance of the algorithm under the case that there are both existing and cold-start developers, 8286 developers are divided into existing developers and cold-start developers according to the ratio of 1: 1. In addition, from the experimental data, we randomly select data corresponding to 80%, 60%, 40%, 20% of the tasks and the existing developers as the training set, denoted as TR80, TR60, TR40, TR20, respectively. We randomly select data corresponding to 10% of the task and all the developers as the validation set, and the data corresponding to the remaining 10%, 30%, 50%, 70% of the task and all the developers as the test set, denoted as TE10, TE30, TE50, TE70, respectively.

In the experiment, we first use both the training set and the validation set to tune the parameters of our model according to MAE, in order to obtain the best performance of our model. Second, the MRFMRec algorithm and the traditional FM algorithm are compared on the test set in terms of *Precision* and *Recall*.

To calculate *Precision* and *Recall* on the test set, we first process the test set: mark developers whose ratings are greater than or equal to 4 as recommended, and developers whose ratings are less than 4 as not recommended. Similarly, when the recommendation system predicts that the rating of cold-start task for a developer is greater than or equal to 4, the developer is considered to be recommended; otherwise, the developer is not considered to be recommended.

On the other hand, we construct the tested developer list $DL = \{d_1, d_2, ..., d_l\}$ on the test set. To ensure that DL includes both recommended and not recommended developers, we first select in the test set those tasks that contain no less than 6 recommended developers and no less than 4 not recommended developers. Second, we randomly select 6 recommended developers and 4 not recommended developers for each task to form the tested developer list.

4.5   Experimental Results and Analysis

Model Tuning. We determine the optimal values of parameters based on the test effect of FM algorithm on the validation set. Since, in the test ranges, there are too many $(8 * 6 * 6 * 6 = 1728)$ combinations for the dimensionality f of latent factor vector and the regularization parameter values λ_{w_0}, λ_w, λ_V, in Table 2, we only give the MAE value corresponding to the optimal combination.

Table 2. The combination of optimal parameters and the corresponding MAE values.

Training set	f	λ_{w_0}	λ_w	λ_V	MAE
TR80	20	0.1	0.1	10	0.881
TR60	25	0.1	1	10	0.902
TR40	25	1	0.1	1	0.939
TR20	20	10	10	1	1.021

Algorithms Comparison. We compare MRFMRec algorithm with the FM algorithm which only considers explicit features. The comparison results (*Precision* and *Recall*) on the test set are shown in Fig. 2.

From Fig. 2, it can be seen that compared with the FM algorithm which only considers explicit features, MRFMRec algorithm has obvious advantages in *Precision* and *Recall*. The experimental results show that the implicit features reflected by the ID information play an important role in improving the performance of the recommendation algorithm.

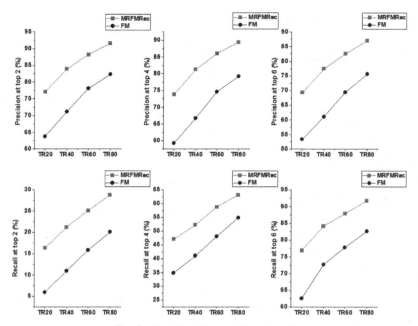

Fig. 2. The *Precision* and *Recall* results.

5 Conclusion

For the cold-start problem of developer recommendation scenarios in crowdsourcing software platform, this paper proposes an FM recommendation algorithm based on explicit to implicit feature mapping relation modeling. The algorithm learns the implicit features of existing tasks and existing developers based on FM algorithm, and learns the mapping relation from explicit features to implicit features based on the deep regression model. Finally, for the cold-start task or developer, the explicit features are used to obtain the implicit features via the regression model, and then the FM model is used to predict ratings by integrating the explicit and implicit features of cold-start tasks or developers. By virtue of the regression model, our model effectively obtains the implicit features of cold-start tasks or developers, and successfully solves the cold-start problem when FM algorithm models the ID information. The comparison experiments on the data set from Topcoder platform show that the proposed algorithm can obtain better recommendation performance by considering both explicit features and ID features.

References

1. Mao, K., Capra, L., Harman, M., Jia, Y.: A survey of the use of crowdsourcing in software engineering. J. Syst. Softw. **126**, 57–84 (2017)
2. Mao, K., Yang, Y., Wang, Q., Jia, Y.: Developer recommendation for crowdsourced software development tasks. In: Service-Oriented System Engineering, pp. 347–356 (2015)

3. Shao, W., Wang, X., Jiao, W.: A developer recommendation framework in software crowd-sourcing development. In: Zhang, L., Xu, C. (eds.) Software Engineering and Methodology for Emerging Domains. CCIS, vol. 675, pp. 151–164. Springer, Singapore (2016). https://doi.org/10.1007/978-981-10-3482-4_11

4. Zhu, J., Shen, B., Hu, F.: A learning to rank framework for developer recommendation in software crowdsourcing. In: Asia-Pacific Software Engineering Conference, pp. 285–292. IEEE, New Delhi (2015)

5. Rendle, S.: Factorization machines with libFM. ACM Trans. Intell. Syst. Technol. **3**(3), 57 (2012)

6. Balabanović, M., Shoham, Y.: Fab: content-based, collaborative recommendation. Commun. ACM **40**(3), 66–72 (1997)

7. Liu, H., Hu, Z., Mian, A.: A new user similarity model to improve the accuracy of collaborative filtering. Knowl. Based Syst. **56**(3), 156–166 (2014)

8. Wan, S., Niu, Z.: A hybrid e-learning recommendation approach based on learners' influence propagation. IEEE Trans. Knowl. Data Eng. **32**(5), 827–840 (2020)

9. He, X., Liao, L., Zhang, H., Nie, L., Hu, X., Chua, T.S.: Neural collaborative filtering. In: The 26th World Wide Web Conference, pp. 173–182. ACM, Perth (2017)

10. Devlin, J., Chang, M.W., Lee, K.: BERT: pre-training of deep bidirectional transformers for language understanding (2018)

11. Vincent, P., Larochelle, H., Lajoie, I.: Stacked denoising autoencoders: learning useful representations in a deep network with a local denoising criterion. J. Mach. Learn. Res. **11**(6), 3371–3408 (2010)

GraphInf: A GCN-based Popularity Prediction System for Short Video Networks

Yuchao Zhang[1]([⊠]), Pengmiao Li[1], Zhili Zhang[2], Chaorui Zhang[3], Wendong Wang[1], Yishuang Ning[4,5], and Bo Lian[6]

[1] Beijing University of Posts and Telecommunications, Beijing, China
zycsmile@163.com
[2] University of Minnesota, Minneapolis, USA
[3] Huawei Company, Hong Kong, China
[4] National Engineering Research Center for Supporting Software of Enterprise Internet Services, Shenzhen 518057, China
[5] Kingdee International Software Group Company Limited, Shenzhen 518057, China
[6] Kuaishou Company, Beijing, China

Abstract. As the emerging entertainment applications, short video platforms, such as Youtube, Kuaishou, quickly dominant the Internet multimedia traffic. The caching problem will surely provide a great reference to network management (e.g., traffic engineering, content delivery). The key to cache is to make precise popularity prediction. However, different from traditional multimedia applications, short video network exposes unique characteristics on popularity prediction due to the explosive video quantity and the mutual impact among these countless videos, making the state-of-the-art solutions invalid. In this paper, we first give an in-depth analysis on 105,231,883 real traces of 12,089,887 videos from *Kuaishou* Company, to disclose the characteristics of short video network. We then propose a graph convolutional neural-based video popularity prediction algorithm called *GraphInf*. In particular, *GraphInf* clusters the countless short videos by region and formulates the problem in a graph-based way, thus addressing the explosive quantity problem. *GraphInf* further models the influence among these regions with a customized graph convolutional neural (GCN) network, to capture video impact. Experimental results show that *GraphInf* outperforms the traditional Graph-based methods by 44.7%. We believe such GCN-based popularity prediction would give a strong reference to related areas.

The work was supported in part by the National Natural Science Foundation of China (NSFC) Youth Science Foundation under Grant 61802024, the Fundamental Research Funds for the Central Universities under Grant 24820202020RC36, the National Key R&D Program of China under Grant 2019YFB1802603, and the CCF-Tencent Rhinoceros Creative Fund under Grant S2019202.

W.-S. Ku et al. (Eds.): ICWS 2020, LNCS 12406, pp. 61–76, 2020.
https://doi.org/10.1007/978-3-030-59618-7_5

1   Introduction

In recent years, online short video (or micro-video) platforms are emerging as a new trend to satisfy the fast-paced modern society. They have been widely spreading all over the world, making video traffic dominate the Internet traffic. As of 2019, there are over 200 million active users in Kuaishou and more than 27 million short videos are being uploaded and viewed, on a daily basis [9].

Video popularity prediction has long been considered as an important topic in Internet traffic area, because it can provide a basis for many network management problems such as caching policies [12,14], reducing the required memory size [28], and modeling videos' lifecycle [27]. Existing popularity prediction algorithms [2,13,19,21,23,24] work well in traditional multimedia scenario, but they become invalid in short video network due to the following two characteristics.

- **Explosive video quantity.** Kuaishou [9] produced more than 5 billion short videos in half a year in 2019, nearly 3,445,900 times more than the total number of TV series (about 351) and films (less than 1100) [26].
- **Relationship among videos.** Online social networks [4] and user behaviour [3,15] play important roles in video popularity, making hot topics propagate from one region to another. Such effects become more apparent in short video network due to its strong social interaction and high timeliness.

Several pioneer efforts have been invested to the short video network prediction problem. [17] uses the average watched percentage of videos to predict the popularity, but becomes inefficient in large scale short video network. [12] takes video content into consideration to predict video popularity, but without considering video relationship, it becomes invalid in short video network. With further research, we made a simple comparison of short videos in different regions. For example, there were 5,755 same videos between Henan in the first 20 min with Anhui in the last 20 min. This shows that videos in different regions influence each other.

In this paper, we propose *GraphInf*, a popularity prediction system in short video network, which is based on a novel customized graph convolutional neural algorithm. *GraphInf* is a highly scalable system that clusters the massive videos into corresponding regions and formulates them by a simple GCN network. The main contributions of this paper are summarized below:

- We disclose the unique characteristics and challenges in short video network by analyzing real data from industry (Sect. 3.1).
- We present a system called *GraphInf* to address the   popularity prediction problem in short video network (Sect. 4).
- We demonstrate the practical benefits of *GraphInf* by building a prototype, and the results also reveal some useful experiences/lessons that would be instructive to related topics. (Sect. 5).

The remainder of this paper is organized as follows. Section 2 briefly reviews the state-of-the-art efforts related to popularity prediction in short video network.

Section 3 introduces the background and motivation of the proposed problem. Section 4 presents the framework of *GraphInf*, with detailed design. Section 5 demonstrates the setting up of *GraphInf* prototype and shows extensive experiment results from real data evaluations. Finally, Sect. 6 concludes this work.

2 Related Work

This work relates to a few areas of active research. We structure our discussion along two parts: the specific characteristics of short video network and video popularity prediction.

2.1 Short Video Network

The specific characteristics of short video network, like large scale and the influence between regions, have not been fully considered yet.

Explosive Video Quantity. It is well known that the number of short videos is huge and the growth rate is quite high. This raises the challenge as to how to cache these videos in limited storage to guarantee the cache hit rate [28]. To reduce the latency due to intermediate caching, [11] proposed a distributed resilient caching algorithm (DR-Cache) that is simple and adaptive to network failures. [25] designed a data store VStore for analytics on large videos. VStore selects video formats catering so as to achieve the target accuracy. [5] designed a parallel processing framework Streaming Video Engine that specially designed to solve the scalability problem, and showed the results of some use cases on data ingestion, parallel processing and overload control.

Relationship Among Videos. The influence between regions depth of short videos is surprising, due to the impact from user behaviour, online social networks, geo-distributed hot events, etc. [7]. [15] proposed an end-to-end framework, DeepInf, which takes user's local network as the input for capturing the latent social representation. [3] discussed the popularity dynamics of videos dynamics of videos in Video Sharing Sites (VSSes) focusing on views, ratings and comments so as to build a emulator which replicates users behaviours in Online Social networks (OSN). These researches focus on the influences between regions model, but didn't make popularity prediction.

In brief, the literature above highlights the special characteristics of short video network that are large scale and the influence between regions.

2.2 Popularity Prediction

Significant efforts have been devoted to exploring item popularity prediction due to the potencial business value [21]. [23] provided the affect the popularity of science communication videos on YouTube. They found that the user-generated contents were significantly more popular than the professionally generated ones and that videos that had consistent science communicators were more popular than those without a regular communicator. [13] proposed LARM that is empowered by a lifetime metric that is both predictable via early-accessible features and adaptable to different observation intervals, as well as a set of specialized regression models to handle different classes of videos with different lifetime. [19] used support vector regression with Gaussian radial basis functions to predict the popularity of an online video measured by its number of views. Although these algorithms have their own advantages, they lack the high performance of operating speed with the rapid growth of the number of short videos.

Overall, we propose a framework *GraphInf* to explore the video popularity in short video network, by taking the special characteristics above into consideration.

3   Motivation

We start by providing some background knowledge of short video network. We point out the uniqueness of such kind of networks by comparing it with traditional online video networks (Sect. 3.1). In particular, by analyzing real statistical figures and access traces from *Kuaishou*, a popular short video platform in China, we disclose its two challenges, the explosive video quantity and the complex relationship among short videos.

We then show the opportunity of solving these two challenges above by mapping the massive short videos into regions and formulating the origin problem into a graph-based problem (Sect. 3.2). The graph structure property motivates us to design the proposed *GraphInf*.

3.1   Characteristics of Short Video Network

In this subsection, we show the differences between traditional online video network and short video network. The differences clearly disclose the unique characteristics (also challenges) of popularity prediction problem in short video network.

Explosive Video Quantity. We introduce the explosive video quantity in two aspects, from the perspective of videos and users, respectively.

– *From the perspective of videos.*

Length of Video Uploaded: In 2018, the length of all the uploaded traditional online videos is about 100 thousands of minutes, while that is 340 millions of minutes in only one short video platform (*Kuaishou*), which is 2,850 times longer [8].

Videos Viewed: In 2018, there are only 46 million views per day in top 10 traditional online video platforms in China [10], while the number is 10 billion in only one short video platform (Toutiao), which is 217 times more than the traditional online videos [18].

– *From the perspective of users.*

Growth Rate of Usage Time: According to QuestMobile's report [16], app usage time of short video grew 521.8% in 2018, while the usage time of online video dropped 12%.

Relationship Among Videos. To show the video relationship from a macro perspective, we draw the effect matrix between each pair of provinces from *Kuaishou*'s real traces. To quantify the impact from province i to province j, we take the number of overlapped videos from province i (in current time slot) and province j (in the next time slot). We show the normalized value in Fig. 1. In the figure, each small cube represents the effect depth from province i to province j. Taking Shanxi province as an example, from the horizontal axis, the column of cubes represents the video effect from Shanxi Province to each of the other provinces, while from the ordinate axis, the row of cubes represents the video effect from each of the other provinces to Shanxi Province. We can find that there is influence between different regions, and the influence between different regions is different. What's more, the effect matrix is time-varying due to the timeliness of short video network.

Traditional popularity prediction approaches are usually based on long-term historical access pattern or fixed relationship, which can not reflect the time-varying impact, and therefore become invalid in short video popularity prediction.

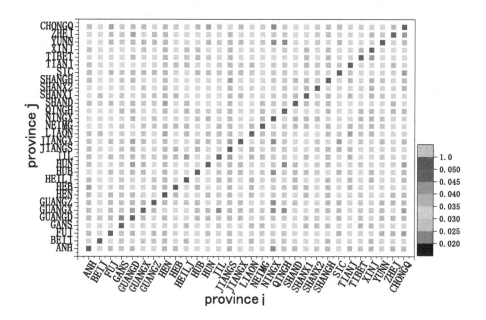

Fig. 1. Popularity effect between any pair of the 31 provinces.

3.2   Potential of Graph-Based Scheme

The characteristics above motivate the need for a lightweight prediction app-roach which can work under huge amount of short videos with influence between regions.

It is known that such relationship among short videos is often modeled by graphs [6,20,22]. Inspired by their success, we model the popularity prediction problem into a graph-based network. Instead of considering each video as a node, we design a geo-distributed clustering scheme to reduce the size of the graph. In particular, we cluster all the short videos into several geo-distributed regions (e.g., provinces/states), and formulate each region as a node in the graph (Fig. 2 shows a simple example). Such graph-based clustering enjoys two significant advantages when predicting video popularity, and therefore has the potential to solve the above two challenges:

– Reduce the calculation scale. Due to the massive amount, it is impractical to predict the popularity for each short video, while such clustering signif-icantly reduces the calculation scale to the number of regions. Therefore,

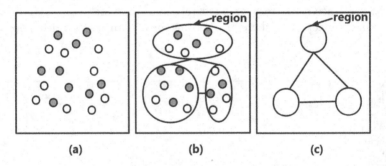

Fig. 2. To predict top n popular videos (red) from N videos (red and white). (a) Predict the popularity for each video. (b) Group the N videos to K regions, and predict top n_i videos in region k_i ($\sum_1^K n_i = N$). (c) Analyze these regions that constitute a simplified graph. (Color figure online)

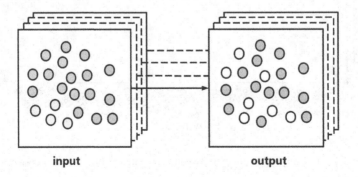

input   output

Fig. 3. The input and output of a *GraphInf*

GraphInf could handle much more short videos and keep the calculation overhead unchanged.
– Get the relationship among videos in different regions. As *GraphInf* formulates the problem into a simple graph network, it thus can get the influence between the short videos (Fig. 2(a)), by updating the attributes of nodes and links.

Overall, in such graph-based formulation, we can model a GCN network which takes all the potential videos as input and predicts the future video popularity, as shown in Fig. 3. Base on the analysis and observations above, the question becomes that how to achieve short video popularity prediction on a graph, with explosive video quantity and time-varying video influence? We therefore design *GraphInf*, which will be introduced in the next section in detail.

4   *GraphInf* Framework

In this section, we first introduce how to formulate the popularity prediction problem of short videos in a graph-based way in Subsect. 4.1. Then, we propose *GraphInf* to deal with the challenges introduced in Sect. 3.1, and describe the components of *GraphInf* in Subsect. 4.2. The notations to be used are shown in Table 1.

4.1   Problem Formulation

Generally, different places have their own popular short videos. The popular short videos of each place are mainly dominated by its local information (e.g., human behavior, age structure) and neighbors. In particular, we distinguish the places by administrative regions, where each region could be a city, a province, or a state. By clustering and analyzing all short videos in the region, we are able to handle the large scale problem of the growing short videos. Then, the connections among all regions are described by a graph $\mathcal{G} = (\mathcal{N}, \mathcal{E})$. Each region is denoted by a node and $\mathcal{N}$ is the set of all nodes. We assume that every two regions are virtually adjacent by using an edge with a weight that represents their relationship. All edges are included in the set $\mathcal{E}$. Next, we formally introduce the node and edge information used here.

Node Information. At each time slice t, every region has its own popular short videos. We divide the short video source into two types of popularity according to request times. One is the popular short videos (PV_{self}) whose request times are over hundreds or even thousands. The other is the sub-popular short videos (PV_{sub}) with a smaller number of request times. Let s_i^t and c_i^t denote the types of the popular and sub-popular short video of region i at time slice t, respectively. Specifically, s_i^t and c_i^t are two vectors with the dimension M, which consists of the request times of short videos. The number M indicates the top M short videos of each type.

Edge Information. Except the geographically adjacent information between two regions, we think that region i is affected by the top short videos of all other regions. Thus, we use the top short video information of all other regions (PV_{other}) as the edge information of region i. Let o_i^t denote the edge information of region i at time slice t. Mathematically, the representation of o_i^t is

$$o_i^t = [s_1^t; ...; s_{i-1}^t; s_{i+1}^t; ...; s_N^t]$$

with dimension $(|\mathcal{N}| - 1) \times M$.

When given the node and edge information of previous $l \geq 1$ time slices, we aim to predict the popularity in the next time slice $t + 1$ in each region. This problem is formally described as below.

$$\{s_i^{t+1}; c_i^{t+1}; o_i^{t+1}\} = f(\{s_i^t; c_i^t; o_i^t\}, \{s_i^{t-1}; c_i^{t-1}; o_i^{t-1}\}, ...,$$
$$\{s_i^{t-l}; c_i^{t-l}; o_i^{t-l}\}), \tag{1}$$

Table 1. Notations

Symbol	Definition
s_i^t	The popular videos of region i at time slice t
c_i^t	The sub-popular videos of region i at time slice t
o_i^t	The top short videos of all other region of region i at time slice t
$w_{i \leftarrow j}^t$	The number of the same short videos appearing in both region j at time slice $t - 1$ and region i at time slice t
W_i^t	$W_i^t = [w_{i \leftarrow 1}^t, ..., w_{i \leftarrow (i-1)}^t, w_{i \leftarrow (i+1)}^t, ..., w_{i \leftarrow N}^t]$
X_i	The importance vector of each feature at region i, $X_i = [x_i^1, x_i^2, x_i^3]$
R_i^{t+1}	The predicted popular videos of region i at time slice $t + 1$
Y_i^{t+1}	The ground truth indicating that the short videos of region i are popular at time slice $t + 1$

where $f(\cdot)$ is the popularity prediction function of short videos. The difficulty of solving problem (1) lies in how to obtain an appropriate graph representation of the effect of hot topic between regions and then use it to overcome the relationship among videos in different regions to get popularity videos in explosive video quantity. For this, we propose *GraphInf* to solve this problem.

4.2   *GraphInf*: A Novel Graph Network

GraphInf is tailored to deal with the challenges mentioned above in short videos. In particular, the architecture of *GraphInf* given in Fig. 4 includes five layers: embedding layer, normalization layer, input layer, GCN layer, and output layer. Initially, we collect the raw data of node and edge information of all nodes at time slice t, i.e., $\{s_i^t; c_i^t; o_i^t\}, \forall i \in \mathcal{N}$. Next, we describe the five steps in turn.

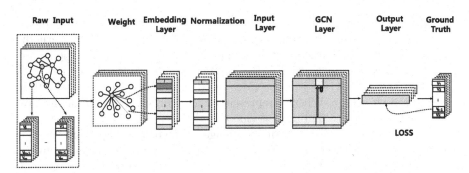

Fig. 4. The architecture of *GraphInf*.

- Embedding layer. The edge information of region i adopted here is the top short videos of all other regions. To quantify the effect of region j to region i at time slice t, we calculate the weight $w^t_{i \leftarrow j}$ to denote the number of the same short videos appearing in both region j at time slice $t-1$ and region i at time slice t. Thus, we use vector W^t_i denoting all weights of all other nodes, i.e.,

$$W^t_i = [w^t_{i \leftarrow 1}, ..., w^t_{i \leftarrow (i-1)}, w^t_{i \leftarrow (i+1)}, ..., w^t_{i \leftarrow \mathcal{N}}]$$

with dimension $|\mathcal{N}| - 1$. (Line 3 in Algorithm 1)
- Normalization layer. After the embedding step, the raw data of region i at time slice t turns to

$$H^t_i = \begin{bmatrix} s^t_i \\ c^t_i \\ W^t_i o^t_i \end{bmatrix}. \tag{2}$$

The entries in H^t_i are the request times of corresponding short videos. To eliminate the magnitude impact, we carry out the normalization of each entry h of H^t_i following Eq. (3). (Line 4 in Algorithm 1)

$$h' = \frac{h - \min(H^t_i)}{\max(H^t_i) - \min(H^t_i)}. \tag{3}$$

- Input layer. In the input layer, we define the normalized term $Normalization(H^t_i)$ as $\hat{H}^t_i$ to be the new input of region i at time slice t. (Line 5 in Algorithm 1)
- GCN layer. We customized GCN layer that independent of the original GCN. The overall operation of the GCN unit for region i follows the next three steps.

$$U^t_i = X_i \cdot \hat{H}^t_i, \tag{4}$$

$$\tilde{U}^t_i = \text{Sort}(U^t_i), \tag{5}$$

$$R^{t+1}_{i,m} = \begin{cases} 1 & \text{if } U^t_{i,m} \geq \tilde{U}^t_{i,k} \\ 0 & \text{otherwise} \end{cases}, \tag{6}$$

Each row of $\hat{H}_i^t$ denotes a pre-defined feature, say the popular, sub-popular, and all other popular short videos of region i. We adopt GCN to learn the importance of each feature at region i. Let $X_i = [x_i^1, x_i^2, x_i^3]$ denote the importance of each feature at region i. $R_{i,m}^{t+1}$ is the predicted outcome and indicates whether the mTH video is a hot video in all short videos, 1 if it is, else 0 ,in region i at time $t+1$. Where k represents a number of hot Videos defined.We use the GCN layer to find the hot videos in the mount of short videos, to deal with the explosive video quantity challenge. (Line 6–8 in Algorithm 1)

– Output layer. In the training process, when we obtain the output from the GCN layer, we have to evaluate the loss between the training result and ground truth so as to update all $X_i, \forall i \in \mathcal{N}$. We define the loss function of each region as follows.

$$loss = \frac{|U_i^t (R_i^{t+1})^T - U_i^t (Y_i^{t+1})^T|^2}{U_i^t (Y_i^{t+1})^T},\tag{7}$$

where Y_i^{t+1} is the ground truth indicating that the K short videos of region i are popular at time slice $t+1$. If yes, the corresponding value in Y_i^{t+1} is set to be 1, otherwise 0. The superscript T is the operation of matrix transpose. (Line 9 in Algorithm 1)

At last, we summarize *GraphInf* in Algorithm 1.

Algorithm 1. The pesudo code of *GraphInf*

---

input: The time sequence training dataset $\mathcal{D}^{train}$ including the video request times $\{s_i^t, c_i^t, o_i^t\}, t \in \mathcal{D}^{train}$ (*Here, the data of the next time slice $t + 1$ is used as the ground truth Y_i^{t+1}*); the initialization $X_i(0)$.

1: **for** t **do**
2:     **for** $i \in \mathcal{N}$ **do**
3:         Calculate W_i^t
4:         Calculate H_i^t
5:         $\hat{H}_i^t = Normalization(H_i^t)$
6:         $U_i^t \leftarrow X_i^t \cdot \hat{H}_i^t$
7:         $U_{i(K)}^t \leftarrow$ The Kth value in Sort(U_i^t)
8:         $R_i^{t+1} \leftarrow$ Find the values in U_i^t that are larger than $U_{i,K}^t$ and denote them as one
9:         Calculate the loss compared with the ground truth Y_i^{t+1} by using equation (7)
10:        **if** $loss^i - loss^{i-1} \leq \epsilon$ **then**
11:            Terminate the algorithm
12:        **end if**
13:        Return $X_i = X_i(t)$ and R_i^{t+1}
14:     **end for**
15: **end for**
output: The importance of features $X_i = [x_i^1, x_i^2, x_i^3]$ and the prediction popular videos R_i^{t+1}

---

5   Evaluation

In this section, we evaluate our approach *GraphInf* using real traces, and show the results of applying *GraphInf* versus the existing representative policies.

5.1   Experiment Setting

Algorithms. We compare *GraphInf* with three representative solutions.

- RNN-based. As popular short videos of each place are mainly dominated by its local information, there are some works that use historical data to predict item popularity, by designing a recurrent network with memory [14]. So we use the historical hot videos to illustrate the potential of RNN-based schemes.
- Graph-based. As we described in the Sect. 3.2, some efficient solutions are modeled by graphs [6,20], so we use the historical sub-hot videos, as comparison of *GraphInf*.
- Embedding. In order to improve the topic prediction accuracy, some works embed specific characteristics of graph node into consideration [1] and achieve more desirable results. We therefore further implement an embedding method as comparison.

Table 2. Information about the dataset.

Dataset	Cache size	Access #.	Server #.	Video #.
Province 1	$3.96T$	$5,323,508$	30	$1,424,564$
Province 2	$5.12T$	$9,202,618$	72	$1,813,058$
Province 3	$2.51T$	$3,051,059$	10	$876,058$
Province 4	$2.39T$	$2,765,419$	21	$843,219$
Province 5	$2.48T$	$2,828,645$	6	$862,806$
...	...	...	...	...
Total	$78.75T$	$105,231,883$	488	$12,089,887$

Datasets. The traces [28] are from 31 provinces with 1,128,989 accesses to 132,722 videos in 1 h (shown in Table 2). Each trace item contains the timestamp, anonymized source IP, video ID and url, file size, location, server ID, cache status, and consumed time (with no personal sensitive information). We then deploy and evaluate *GraphInf* by comparing with representative algorithms.

5.2   System Performance

We first conduct a series of experiments to show the overall prediction performance. In particular, we show the popular video prediction accuracy in each region, and then deeply look into these popular videos by analyzing their source and the prediction accuracy correspondingly.

Overall Accuracy. As described in problem formulation section (Sect. 4.1), we consider both node information (popular videos PV_{self} and sub popular videos PV_{sub}) and edge information (popular videos from other regions PV_{other}). Here we divide them into: the top popular 300 videos from the same region, the top 301 to 600 videos from the same region, and the top popular 300 videos from other 30 regions, respectively. We use this data as input for *GraphInf*, and the output is a matrix of hot videos.

Figure 5 shows the average prediction accuracy in 19 min of all the provinces in China. Figure 6 and 7 show the prediction accuracy comparison of ShanXi and Tibet province in 19 min, respectively. The reason why we choose these two provinces is that ShanXi is a populous province with 10 times the population but only 12% area compared with the sparse Tibet (about 124 times the population density). From these three figures, we can see that *GraphInf* exceeds the other three methods in a populous province (Fig. 6). Surprisingly, we thought it would be easier to predict hot topics in the sparse provinces because its data size is relatively small and the topology is also simpler, but the results show that the accuracy in Tibet is unexpectedly low. In order to figure this out, we further conduct experiments to calculate the specific accuracy by video source (PV_{self}, PV_{sub} and PV_{other}).

Source Accuracy. To analyze the power of *GraphInf* in detail, here we differentiate these three video sources and check the source accuracy separately. So in each experiment results, there are three bars, denoting the source accuracy from video source PV_{self}, PV_{sub}, and PV_{other}, respectively. The results are shown in Fig. 8. There are more than 2,500 videos (1 min) to get 300 hot videos. The up 19 series of experiments illustrate the average source accuracy of all the 31 cities (in 19 min), the middle 19 series of experiments show the results

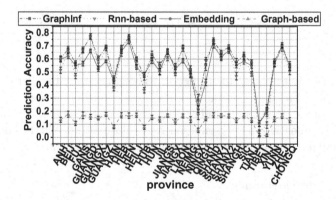

Fig. 5. Prediction accuracy of the four algorithms, in all the 31 cities.

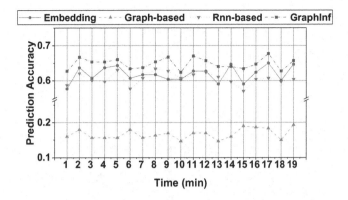

Fig. 6. Prediction accuracy comparison in ShanXi Province, during 19 min.

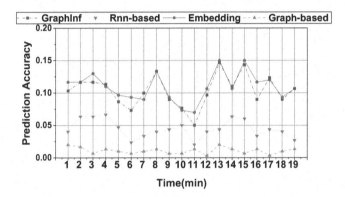

Fig. 7. Prediction accuracy comparison in Tibet Province, during 14 min.

of ShanXi province, and the down 19 series of experiments show the results of Tibet province. From these results we can see that Tibet is lower (any bar) than the average (in the left 19 series experiments), but Henan is higher than the average. So, we can get a result that we obtain a high accuracy of data sources as input in popular province, and a low accuracy in sparse province. The reason for this phenomenon is that it is hard to formulate the video from those sparse provinces, mainly due to the relatively strong closure. Therefore, compared with other cities, the prediction accuracy decreases significantly.

To summarize, the relationship between hot videos in different regions will have an impact on future popularity. *GraphInf* uses a graph to model different regions so as to extract the influence of hot videos in these regions, that's why *GraphInf* can get higher prediction accuracy when compared with existing algorithms. The prediction accuracy of our algorithm will be further improved in the conditions with strong population mobility or close region relationship.

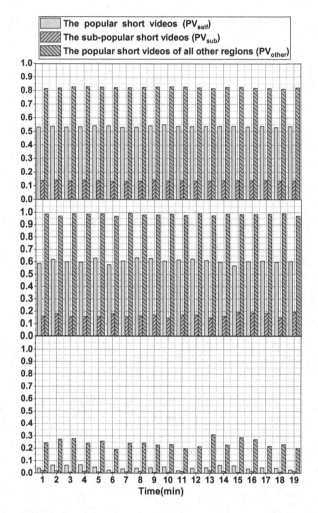

Fig. 8. [Sources accuracy] The sources accuracy w.r.t. different video sources (PV_{self}, PV_{sub} and PV_{other}). Up: the average accuracy of 31 cities. Middle: accuracy of ShanXi Province. Down: acuuracy of Tibet Province.

6   Conclusion

In this work, we shed light on the popularity prediction problem in short video network. We first disclose the specific characteristics of such network from online application data, i.e., explosive video quantity and relationship among videos. Based on the observations, we formulate this problem to a graph and propose a graph-based neural network by incorporating network embedding, normalization and graph convolution. We evaluate the proposed *GraphInf* using real online traces from *Kuaishou* Company and compare the results with three state-of-the-art methods. Experimental results show that *GraphInf* significantly outperforms the baselines with precise topic prediction in short video network.

References

1. Battaglia, P.W., et al.: Relational inductive biases, deep learning, and graph networks. arXiv preprint arXiv:1806.01261 (2018)
2. Chen, J., Song, X., Nie, L., Wang, X., Zhang, H., Chua, T.S.: Micro tells macro: predicting the popularity of micro-videos via a transductive model. In: Proceedings of the 24th ACM International Conference on Multimedia, pp. 898–907. ACM (2016)
3. Ghosh, S., Kumar, S.: Video popularity distribution and propagation in social networks. Int. J. Emerg. Trends Technol. Comput. Sci. (IJETTCS) 6(1), 1–5 (2017)
4. Huang, C., Li, J., Ross, K.W.: Can internet video-on-demand be profitable? In: ACM SIGCOMM Computer Communication Review, vol. 37, pp. 133–144. ACM (2007)
5. Huang, Q., et al.: SVE: distributed video processing at Facebook scale, pp. 87–103, October 2017
6. Hussein, N., Gavves, E., Smeulders, A.W.: Videograph: recognizing minutes-long human activities in videos. arXiv preprint arXiv:1905.05143 (2019)
7. Jampani, V., Gadde, R., Gehler, P.V.: Video propagation networks. In: Proceedings of the IEEE Conference on Computer Vision and Pattern Recognition, pp. 451–461 (2017)
8. Jianshu: Analysis report of tiktok (2018). https://www.jianshu.com/p/2097f2dda7b0
9. kuaishou: Kuaishou (2019). https://www.kuaishou.com
10. lbzuo: The top 10 popular video websites in china (2018). http://www.lbzuo.com/ziyuan/show-15766.html
11. Li, J., Phan, T.K., Chai, W.K., Tuncer, D., Rio, M.: Dr-cache: distributed resilient caching with latency guarantees. In: IEEE INFOCOM (2018)
12. Li, S., Xu, J., Van Der Schaar, M., Li, W.: Popularity-driven content caching. In: IEEE INFOCOM 2016-The 35th Annual IEEE International Conference on Computer Communications, pp. 1–9. IEEE (2016)
13. Ma, C., Yan, Z., Chen, C.W.: Larm:a lifetime aware regression model for predicting YouTube video popularity. In: Proceedings of the 2017 ACM on CIKM, pp. 467–476. ACM (2017)
14. Narayanan, A., Verma, S., Ramadan, E., Babaie, P., Zhang, Z.L.: Deepcache: a deep learning based framework for content caching. In: Proceedings of the 2018 Workshop on Network Meets AI & ML, pp. 48–53. ACM (2018)
15. Qiu, J., Tang, J., Ma, H., Dong, Y., Wang, K., Tang, J.: Deepinf: social influence prediction with deep learning. In: Proceedings of the 24th ACM SIGKDD International Conference on Knowledge Discovery & Data Mining, pp. 2110–2119. ACM (2018)
16. QuestMobile: 2018 china mobile internet (2019). http://www.questmobile.com.cn/research/report-new/29
17. Tan, Z., Zhang, Y., Hu, W.:Online prediction of video popularity in OVSS: a video age-sensitive model with beyond views features. IEEE Trans. Broadcast. pp. 1–10 (2019)
18. Toutiao (2019). https://www.toutiao.com
19. Trzciski, T., Rokita, P.: Predicting popularity of online videos using support vector regression. IEEE Trans. Multimedia 19(11), 2561–2570 (2017)

20. Tsai, Y.H.H., Divvala, S., Morency, L.P., Salakhutdinov, R., Farhadi, A.: Video relationship reasoning using gated spatio-temporal energy graph. In: Proceedings of the IEEE Conference on Computer Vision and Pattern Recognition, pp. 10424–10433 (2019)
21. Vasconcelos, M., Almeida, J.M., Gonçalves, M.A.: Predicting the popularity of micro-reviews: a foursquare case study. Inf. Sci. **325**, 355–374 (2015)
22. Wang, X., Gupta, A.: Videos as space-time region graphs. In: Ferrari, V., Hebert, M., Sminchisescu, C., Weiss, Y. (eds.) ECCV 2018. LNCS, vol. 11209, pp. 413–431. Springer, Cham (2018). https://doi.org/10.1007/978-3-030-01228-1_25
23. Welbourne, D.J., Grant, W.J.: Science communication on YouTube: factors that affect channel and video popularity. Publ. Underst. Sci. **25**(6), 706–718 (2016)
24. Wu, B., Shen, H.: Analyzing and predicting news popularity on Twitter. Int. J. Inf. Manage. **35**(6), 702–711 (2015)
25. Xu, T., Botelho, L.M., Lin, F.X.: Vstore: a data store for analytics on large videos. CoRR abs/1810.01794 (2018)
26. Xueqiu (2020). https://xueqiu.com/9231373161/136685735
27. Yu, H., Xie, L., Sanner, S.: The Lifecyle of a YouTube video: phases, content and popularity. In: Ninth International AAAI Conference on Web and Social Media. (2015)
28. Yuchao, Z., et al.: Challenges and chances for the emerging short video network. In: IEEE International Conference on Computer Communications (Infocom). IEEE (2019)

Reducing the Cost of Aggregation in Crowdsourcing

Rituraj Singh[✉], Loïc Hélouët, and Zoltan Miklos

Univ. Rennes/INRIA/CNRS/IRISA, Rennes, France
{rituraj.singh,zotlan.miklos}@irisa.fr, loic.helouet@inria.fr

Abstract. Crowdsourcing is a way to solve problems that need human contribution. Crowdsourcing platforms distribute replicated tasks to workers, pay them for their contribution, and aggregate answers to produce a reliable conclusion. A fundamental problem is to infer a correct answer from the set of returned results. Another challenge is to obtain a reliable answer at a reasonable cost: unlimited budget allows hiring experts or large pools of workers for each task but a limited budget forces to use resources at best.

This paper considers crowdsourcing of simple boolean tasks. We first define a probabilistic inference technique, that considers difficulty of tasks and expertise of workers when aggregating answers. We then propose CrowdInc, a greedy algorithm that reduces the cost needed to reach a consensual answer. CrowdInc distributes resources dynamically to tasks according to their difficulty. We show on several benchmarks that CrowdInc achieves good accuracy, reduces costs, and we compare its performance to existing solutions.

1 Introduction

Crowdsourcing is a way to solve tasks that need human contribution. These tasks include image annotation or classification, polling, etc. Employers publish tasks on an Internet platform, and these tasks are realized by workers in exchange for a small incentive [2]. Workers are very heterogeneous: they have different origins, domains of expertise, and expertise levels. One can even consider malicious workers, that return wrong answers on purpose. To deal with this heterogeneity, tasks are usually replicated: each task is assigned to a *set of workers*. Redundancy is also essential to collect workers opinion: in this setting, work units are the basic elements of a larger task that can be seen as a poll. One can safely consider that each worker executes his assigned task independently, and hence returns his own belief about the answer. As workers can disagree, the role of a platform is then to build a consensual final answer out of the values returned. A natural way to derive a final answer is **Majority Voting** (MV), i.e. choose as conclusion the most represented answer. A limitation of MV is that all answers have equal weight, regardless of expertise of workers. If a crowd is composed of only few experts, and of a large majority of novices, MV favors answers from

© Springer Nature Switzerland AG 2020
W.-S. Ku et al. (Eds.): ICWS 2020, LNCS 12406, pp. 77–95, 2020.
https://doi.org/10.1007/978-3-030-59618-7_6

novices. However, in some domains, an expert worker may give better answer than a novice and his answer should be given more weight. One can easily replace MV by a weighted vote. However, this raises the question of measuring workers expertise, especially when workers competences are not known a priori.

Crowdsourcing platforms such as Amazon Mechanical Turk (AMT) do not have prior knowledge about the expertise of their worker. A way to obtain initial measure of workers expertise is to use **Golden Questions** [9]. Several tasks with known *ground truth* are used explicitly or hidden to evaluate workers expertise. As already mentioned, a single answer for a particular task is often not sufficient to obtain a reliable answer, and one has to rely on redundancy, i.e. distribute the same task to several workers and aggregate results to build a final answer. Standard *static* approaches on crowdsourcing platforms fixprior number of k workers per task. Each task is published on the platform and waits for bids by k workers. There is no guideline to set the value for k, but two standard situations where k is fixed are frequently met. The first case is when a client has n tasks to complete with a total budget of B_0 incentive units. Each task can be realized by $k = B_0/n$ workers. The second case is when an initial budget is not known, and the platform fixes an arbitrary redundancy level. In this case, the number of workers allocated to each task is usually between 3 and 10 [7]. It is assumed that the distribution of work is uniform, i.e. that each task is assigned the same number of workers. An obvious drawback of static allocation of workers is that all tasks benefit from the same work power, regardless of their difficulty. Even a simple question where the variance of answers is high calls for sampling of larger size. So, one could expect each task t to be realized by k_t workers, where k_t is a number that guarantee that the likelihood to change the final answer with one additional worker is low. However, without prior knowledge on task's difficulty and on variance in answers, this number k_t cannot be fixed.

This paper proposes a new algorithm called CrowdInc to address the questions of answers aggregation, task allocation, and cost of crowdsourcing. For simplicity, we consider boolean filtering tasks, i.e. tasks with answers in $\{0, 1\}$, but the setting can be easily extended to tasks with any finite set of answers. These tasks are frequent, for instance to decide whether a particular image belongs or not to a given category of pictures. We consider that each binary task has a *truth label*, i.e. there exists a ground truth for each task. Each worker is asked to answer 0 or 1 to such a task and returns a so-called *observed label*, which may differ from the ground truth. The *difficulty* of a task is a real value in $[0, 1]$. A task with difficulty 0 is a very easy task and a task with difficulty 1 a very complex one. The *expertise* of a worker is modeled in terms of *recall* and *specificity*. **Recall** (also called true positive rate) measures the proportion of correct observed labels given by a worker when the ground truth is 1. On contrary, **specificity** (also called true negative rate) measures the proportion of correct observed labels given by a worker when the ground truth is 0. We propose a generating function to measure the probability of accuracy for each of the truth label (0/1) based on the *observed label, task difficulty, and worker expertise*. We rely on an Expectation Maximization (EM) based algorithm to maximize the probability of accuracy of ground truth for each task and jointly estimate the difficulty of each task as well as expertise of the workers. The algorithm provides a greater weight to expert workers. In addition, if a worker with high *recall* makes a mistake in the *observed label*, then it increases the difficulty of the task (correspondingly for specificity). Along with, if expert workers fail to return a correct answer, then the task is considered difficult. The EM algorithm converges with a very low error rate and at the end returns the task *difficulty*, worker *expertise* and the *final estimated label* for each task based on *observed labels*. Additionally, we propose a dynamic worker allocation algorithm that handles at the same time aggregation of answers, and optimal allocation of a budget to reach a consensus among workers. The algorithm works in two phases. For the initial *Estimation* phase, as we do not have any prior information about the task difficulty and worker expertise, we allocate one third of total budget to inspect the behavior of each task. Based on the answers provided by the human workers for each task, we first derive the difficulty of tasks, final aggregated answers, along with the worker expertise using an EM algorithm. For each task, we estimate the likelihood that the aggregated answer is the ground truth. We terminate tasks which are above the derived threshold at that particular instance. The second phase is an *Exploration* phase. Based on each of the estimated task difficulty, we start to allocate workers for each of the remaining tasks. The process continues until all tasks are terminated or the whole budget is consumed.

Related Work: Several papers have considered tools such as EM to aggregate answers, or allocate tasks. We only highlight a few works that are close to our approach, and refer interested readers to [20] for a more complete survey of the domain. Zencrowd [4] considers workers competences in terms of accuracy (ratio of correct answers) and aggregates answers using EM. PM [10] considers an optimization scheme based on Lagrange multipliers. Workers accuracy and ground truth are the hidden variables that must be discovered in order to minimize the deviations between workers answers and aggregated conclusion. D&S [3] uses EM to synthesize answers that minimize error rates from a set of patient records. It considers recall and specificity, but not difficulty of tasks. The approach of [8] proposes an algorithm to assign tasks to workers, synthesize answers, and reduce the cost of crowdsourcing. It assumes that all tasks have the same difficulty, and that worker reliability is a consistent value in [0, 1] (hence considering accuracy as a representation of competences). CrowdBudget [15] is an approach that divides a budget B among K existing tasks to achieve a low error rate, and then uses MV to aggregate answers. Workers answers follow an unknown Bernoulli distribution. The objective is to affect the most appropriate number of workers to each task in order to reduce the estimation error. Aggregation is done using Bayesian classifiers combination (BCC). The approach in [16] extends BCC with communities and is called CBCC. Each worker is supposed to belong to a particular (unknown) community, and to share characteristics of this community (same recall and specificity). This assumption helps improving accuracy of classification. Expectation maximization is used by [14] to improve supervised learning when the ground truth in unknown. This work considers recall and specificity of workers and proposes maximum-likelihood estimator that jointly learns a classifier, discovers the best experts, and estimates ground truth. Most of the works cited above consider expertise of workers but do not address tasks difficulty. An exception is GLAD (Generative model of Labels, Abilities, and Difficulties) [19] that proposes to estimate tasks difficulty as well as workers accuracy to aggregate final answers. The authors recall that EM is an iterative process that stops only after converging, but demonstrate that the EM approach needs only a few minutes to tag a database with 1 million images. The authors in [1] consider difficulty and error parameter of the worker. Notice that in most of the works, tasks difficulty is not considered and expertise is modeled in terms of accuracy rather than recall and specificity. Generally the database and machine learning communities focus on data aggregation techniques and leave budget optimization apart. Raykar et al. [13] introduce sequential crowdsourced labeling: instead of asking for all the labels in one shot, one decides at each step whether evaluation of a task shall be stopped, and which worker should be hired. The model incorporates a Bayesian model for workers (workers are only characterized by their accuracy), and cost. Then, sequential crowdsourced labeling amounts to exploring a (very large) Markov decision process (states contain all pairs of task/label collected at a given instant) with a greedy strategy.

It is usually admitted [20] that recall and specificity give a finer picture of worker's competence than accuracy. Our work aggregates workers answers using expectation maximization with three parameters: task difficulty, recall and specificity of workers. The CrowdInc algorithm uses this EM aggregation to estimate error and difficulty of tasks. This error allows to compute dynamically a threshold to stop tasks which aggregated answers have reached a reasonable reliability and to allocate more workers to the most difficult tasks, hence saving costs. One can notice that we assign an identical cost to all tasks. This makes sense, as the difficulty of tasks is initially unknown.

The rest of the paper is organized as follows. In Sect. 2, we introduce our notations, the factors that influence results during aggregation of answers, and the EM algorithm. In Sect. 3, we present a model for workers and our EM-based aggregation technique. We detail the CrowdInc algorithm to optimize the cost of crowdsourcing in Sect. 4. We then give results of experiments with our aggregation technique and with CrowdInc in Sect. 5. Finally we conclude and give future research directions in Sect. 6.

2   Preliminaries

In the rest of the paper, we will work with discrete variables and discrete probabilities. A *random variable* is a variable whose value depends on random phenomenon. For a given variable x, we denote by $Dom(x)$ its domain (boolean, integer, real, string, ...). For a particular value $v \in Dom(x)$ we denote by $x = v$ the event "x has value v". A probability measure $Pr()$ is a function from a domain to interval $[0, 1]$. We denote by $Pr(x = v)$ the probability that event $x = v$ occurs. In the rest of the paper, we mainly consider boolean events, i.e.

variables with domain $\{0,1\}$. A probability of the form $Pr(x = v)$ only considers occurrence of a single event. When considering several events, we define the *joint probability* $Pr(x = v, y = v')$ the probability that the two events occur simultaneously. The notation extends to an arbitrary number of variables. If x and y are independent variables, then $Pr(x = v, y = v') = Pr(x = v) \cdot Pr(y = v')$. Last, we will use conditional probabilities of the form $Pr(x = v \mid y = v')$, that defines the probability for an event $x = v$ when it is known that $y = v'$. We recall that, when $P(y = v') > 0$ $Pr(x = v \mid y = v') = \frac{Pr(x=v, y=v')}{Pr(y=v')}$.

2.1   Factors Influencing Efficiency of Crowdsourcing

During task labeling, several factors can influence the efficiency of crowdsourcing, and the accuracy of aggregated answers. The first one is **Task difficulty**. Tasks submitted to crowdsourcing platforms by a client address simple questions, but may nevertheless require some expertise. Even within a single application type, the difficulty for the realization of a particular task may vary from one experiment to another: tagging an image can be pretty simple if the worker only has to decide whether the picture contains an animal or an object, or conversely very difficult if the boolean question asks whether a particular insect picture shows an hymenopteran (an order of insects). Similarly, **Expertise of workers** plays a major role in accuracy of aggregated answers. In general, an expert worker performs better on a specialized task than a randomly chosen worker without particular competence in the domain. For example, an entomologist can annotate an insect image more precisely than any random worker.

The technique used for **Amalgamation** also play a major role. Given a set of answers returned for a task t, one can aggregate the results using *majority voting* (MV), or more interesting, as a weighted average answer where individual answers are pondered by workers expertise. However, it is difficult to get a prior measure of workers expertise and of the difficulty of tasks. Many crowdsourcing platforms use MV and ignore difficulty of tasks and expertise of workers to aggregate answers or assign tasks to workers. We show in Sect. 5 that MV has a low accuracy. In our approach, expertise and difficulty are hidden parameters evaluated from the sets of answers returned. This allows considering new workers with a priori unknown expertise. One can also start with an a priori measure of tasks difficulty and of workers expertise. Workers expertise can be known from former interactions. It is more difficult to have an initial knowledge of tasks difficulties, but one can start with an a priori estimation. However, these measures need to be re-evaluated on the fly when new answers are provided by the crowd. Starting with a priori measures does not change the algorithms proposed hereafter, but may affect the final aggregated results.

In Sect. 3, we propose a technique to estimate the expertise of workers and difficulty of tasks on the fly. Intuitively, one wants to consider a task difficult if even experts fail to provide a correct answer for this task, and consider it easy if even workers with low competence level answer correctly. Similarly, a worker is competent if he answers correctly difficult tasks. Notice however that to measure difficulty of tasks and expertise of workers, one needs to have the final answer for each task. Conversely, to precisely estimate the final answer one needs to have the worker expertise and task difficulty. This is a chicken and egg situation, but we show in Sect. 3 how to get plausible value for both using EM.

The next issue to consider is the **cost** of crowdsourcing. Workers receive incentives for their work, but usually clients have limited budgets. Some task may require a lot of answers to reach a consensus, while some may require only a few answers. Therefore, a challenge is to spend efficiently the budget to get the most accurate answers. In Sect. 4, we discuss some of the key factors in budget allocation. Many crowdsourcing platforms do not considers *difficulty*, and allocate the same number of workers to each task. The allocation of many workers to simple tasks is usually not justified and is a waste of budget that would be useful for difficult tasks. Now, tasks difficulty is not a priori known. This advocates for on the fly worker allocation once the difficulty of a task can be estimated. Last, one can stop collecting answers for a task when there is an evidence that enough answers have been collected to reach a consensus on a final answer. A immediate solution is to measure the confidence of final aggregated answer and take as **Stopping Criterion** for a task the fact that this confidence exceeds a chosen threshold. However, this criterion does not works well in practice as clients usually want high thresholds for all their tasks. This may lead to consuming all available budget without reaching an optimal accuracy. Ideally, we would like to have a stopping criterion that balances confidence in the final answers and budget, and optimizes the overall accuracy of answers for all the tasks.

2.2 Expectation Maximization

Expectation Maximization [5] is an iterative technique to obtain maximum likelihood estimation of parameter of a statistical model when some parameters are unobserved and *latent*, i.e. they are not directly observed but rather inferred from observed variables. In some sense, the EM algorithm is a way to find the best fit between data samples and parameters. It has many applications in machine learning, data mining and Bayesian statistics.

Let $\mathcal{M}$ be a model which generates a set $\mathcal{X}$ of observed data, a set of missing latent data $\mathcal{Y}$, and a vector of unknown parameters θ, along with a likelihood function $L(\theta \mid \mathcal{X}, \mathcal{Y}) = p(\mathcal{X}, \mathcal{Y} \mid \theta)$. In this paper, observed data $\mathcal{X}$ represents the answers provided by the crowd, $\mathcal{Y}$ depicts the *final answers* which need to be estimated and are hidden, and parameters in θ are the *difficulty* of tasks and the *expertise* of workers. The *maximum likelihood estimate* (MLE) of the unknown parameters is determined by maximizing the marginal likelihood of the observed data. We have $L(\theta \mid \mathcal{X}) = p(\mathcal{X} \mid \theta) = \int p(\mathcal{X}, \mathcal{Y} \mid \theta) d\mathcal{Y}$. The EM algorithm computes iteratively MLE, and proceeds in two steps. At the kth iteration of the algorithm, we let θ^k denote the estimate of parameters θ. At the first iteration of the algorithm, θ^0 is randomly chosen.

E-Step: In the E step, the missing data are estimated given observed data and current estimate of parameters. The E-step computes the expected value of

$L(\theta \mid \mathcal{X}, \mathcal{Y})$ given the observed data $\mathcal{X}$ and the current parameter θ^k. We define

$$Q(\theta \mid \theta^k) = \mathbb{E}_{\mathcal{Y}\mid\mathcal{X},\theta^k}[L(\theta \mid \mathcal{X}, \mathcal{Y})] \qquad (1)$$

In the crowdsourcing context, we use the E-Step to compute the probability of occurrence of $\mathcal{Y}$ that is the *final answer* for each task, given the observed data $\mathcal{X}$ and parameters θ^k obtained at k^{th} iteration.

M-Step: The M-step finds parameters θ that maximize the expectation computed in Eq. 1.

$$\theta^{k+1} = \arg\max_{\theta} Q(\theta \mid \theta^k) \qquad (2)$$

Here, with respect to estimated probability for $\mathcal{Y}$ for *final answers* from the last E-Step, we maximize the joint log likelihood of the observed data $\mathcal{X}$ (answer provided by the crowd), hidden data $\mathcal{Y}$ (final answers), to estimate the new value of θ^{k+1} i.e. the *difficulty* of tasks and the *expertise* of workers. The E and M steps are repeated until the value of θ^k converges. A more general version of the algorithm is presented in Algorithm 1.

Algorithm 1: General EM Algorithm

    Data: Observed Data $\mathcal{X}$
    Result: Parameter values θ, Hidden data $\mathcal{Y}$
1 Initialize parameters in θ^0 to some random values.
2 **while** $||\theta^k - \theta^{k-1}|| > \epsilon$ **do**
3 | Compute the expected possible value of $\mathcal{Y}$, given θ^k and observed data $\mathcal{X}$
4 | Use $\mathcal{Y}$ to compute the values of θ that maximize $Q(\theta \mid \theta^k)$.
5 **end**
6 return parameter θ^k, Hidden data $\mathcal{Y}$

---

3   The Aggregation Model

We address the problem of evaluation of binary properties of samples in a dataset by aggregation of answers returned by participants in a crowdsourcing system. This type of application is frequently met: one can consider for instance a database of n images, for which workers have to decide whether each image is clear or blur, whether a cat appears on the image, etc. The evaluated property is binary, i.e. workers answers can be represented as a label in $\{0,1\}$. From now, we will consider that tasks are elementary work units which objective is to associate a binary label to a particular input object. For each task, an actual ground truth exists, but it is not known by the system. We assume a set of k independent workers, which role is to realize a task, i.e. return an *observed label* in $\{0,1\}$ according to their perception of a particular sample. We consider a set of tasks $T = \{t_1, \ldots t_n\}$ for which a label must be evaluated. For a task $t_j \in T$ the observed label given by worker $1 \le i \le k$ is denoted by l_{ij}. We let y_j denote the *final label* of a task t_j obtained by aggregating the answers of all workers.

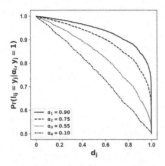 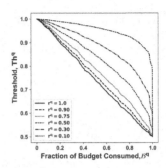

Fig. 1. (left) Generative function for the probability to get $l_{ij} = 1$, given $y_j = 1$, for growing values of task difficulty. The curves represent different recall for the considered workers. (right) The threshold values based on current estimate on consumed budget and fraction of task remaining at the beginning of a round.

$L_j = \bigcup_{i \in 1..k} l_{ij}$ denotes the set of all labels returned by workers for task t_j, L denotes the set of all observed labels, $L = \bigcup_{j \in 1..n} L_j$. The goal is to estimate the ground truth by synthesizing a set of *final label* $Y = \{y_j, 1 \le j \le n\}$ from the set of *observed label* $L = \{L_j\}$ for all tasks.

Despite the apparent simplicity of the problem, crowdsourcing binary tagging tasks hides several difficulties, originating from unknown parameters. These parameters are the difficulty of each task, and the expertise of each worker. The difficulty of task t_j is modeled by a parameter $d_j \in (0,1)$. Here value 0 means that the task is very easy, and can be performed successfully by any worker. On the other hand, $d_j = 1$ means that task t_j is very difficult. A standard way to measure expertise is to define workers accuracy as a pair $\xi_i = \{\alpha_i, \beta_i\}$, where α_i is called the *recall* of worker i and β_i the *specificity* of worker i. The **recall** is the probability that worker i annotates an image j with label 1 when the ground truth is 1, i.e. $\alpha_i = Pr(l_{ij} = 1 | y_j = 1)$. The **specificity** of worker i is the probability that worker i annotates an image j with 0 when the ground truth is 0, i.e. $\beta_i = Pr(l_{ij} = 0 | y_j = 0)$.

In literature, [20] the expertise of workers is often quantified in terms of *accuracy*, i.e. $Pr(l_{ij} = y_j)$. However, if the data samples are unbalanced, i.e. the number of samples with actual ground truth 1 (respectively 0) is much larger than the number of samples with ground truth 0 (respectively 1), defining competences in terms of *accuracy* leads to bias. Indeed, a worker who is good in classifying images with ground truth 1 can obtain bad scores when classifying image with ground truth 0, and yet get a good accuracy (this can be the case of a worker that always answers 1 when tagging a task). *Recall* and *Specificity* overcomes the problem of bias and separates the worker expertise, considering their ability to answer correctly when the ground truth is 0 and when it is 1, and hence give a more precise representation of workers competences.

Recall and specificity allows us to build a probabilistic model (a generative model) for workers answers. We assume that workers have constant behaviors and are faithful, i.e. do not return wrong answers intentionally. We also assume that workers do not collaborate (their answers are independent variables). Under these assumptions, knowing the recall α_i and specificity β_i of a worker i, we build a model that generates the probability that he returns an *observed label* l_{ij} for a task j with difficulty d_j:

$$Pr(l_{ij} = y_j | d_j, \alpha_i, y_j = 1) = \frac{1 + (1 - d_j)^{(1-\alpha_i)}}{2} \tag{3}$$

$$Pr(l_{ij} = y_j | d_j, \beta_i, y_j = 0) = \frac{1 + (1 - d_j)^{(1-\beta_i)}}{2} \tag{4}$$

Figure 1-(left) shows the probability of associating label 1 to a task for which the ground truth is 1 when the difficulty of the tagging task varies, and for different values of recall. The range of task difficulty is $[0, 1]$. The vertical axis is the probability of getting $l_{ij} = 1$. One can notice that this probability takes values between 0.5 and 1. Indeed, if a task is too difficult, then returning a value is close to making a random guess of a binary value. Unsurprisingly, as the difficulty of task increases, the probability of correctly labeling the task decreases. This generative function applies for every worker. For a fixed difficulty of task, workers with higher recalls have higher probability to correctly label a task. Also, note that when the difficulty of a task approaches 1, the probability of answering with label $l_{ij} = 1$ decreases for every value of α_j. However, for workers with high recall, the probability of a correct annotation is always greater than with a smaller recall. Hence, the probability of correct answer depends both on the difficulty of task and on expertise of the worker realizing the task.

3.1   Aggregating Answers

For a given task j, with unknown difficulty d_j, the answers returned by k workers (observed data) is a set $L_j = \{l_{1j}, \ldots, l_{kj}\}$, where l_{ij} is the answer of worker i to task j. In addition, workers expertise are vectors of parameters $\alpha = \{\alpha_1, \ldots \alpha_k\}$ and $\beta = \{\beta_1, \ldots \beta_k\}$ and are also unknown. The goal is to infer the final label y_j, and to derive the most probable values for d_j, α_i, β_i, given the observed answers of workers. We use a standard EM approach to infer the most probable actual answer $Y = \{y_1, \ldots y_n\}$ along with the hidden parameters $\Theta = \{d_j, \alpha_i, \beta_i\}$. Let us consider the E and M phases of the algorithm.

E Step: We assume that all answers in $L = \bigcup_{1 \leq j \leq k} L_j$ are independently given by the workers as there is no collaboration between them. So, in every $L_j = \{l_{1j}, \ldots, l_{kj}\}$, l_{ij}'s are independently sampled variables. We compute the posterior probability of $y_j \in \{0, 1\}$ for a given task j given the difficulty of task d_j, worker expertise $\alpha_i, \beta_i, i \leq k$ and the worker answers $L_j = \{l_{ij} \mid i \in 1..k\}$. Using Bayes' theorem, considering a particular value $\lambda \in \{0, 1\}$ we have:

$$Pr[y_j = \lambda | L_j, \alpha, \beta, d_j] = \frac{Pr(L_j | y_j = \lambda, \alpha, \beta, d_j) \cdot Pr(y_j = \lambda | \alpha, \beta, d_j)}{Pr(L_j \mid \alpha, \beta, d_j)} \tag{5}$$

One can remark that y_j and α, β, d_j are independent variables. We assume that both values of y_j are equiprobable, i.e. $Pr(y_j = 0) = Pr(y_j = 1) = \frac{1}{2}$. We hence get:

$$Pr[y_j = \lambda | L_j, \alpha, \beta, d_j] = \frac{Pr(L_j | y_j = \lambda, \alpha, \beta, d_j) \cdot Pr(y_j = \lambda)}{Pr(L_j | \alpha, \beta, d_j)} = \frac{Pr(L_j | y_j = \lambda, \alpha, \beta, d_j) \cdot \frac{1}{2}}{Pr(L_j | \alpha, \beta, d_j)} \quad (6)$$

Similarly, the probability to obtain a particular set of labels is given by:

$$Pr(L_j \mid \alpha, \beta, d_j) = \frac{1}{2} \cdot Pr(L_j \mid y_j = 0, \alpha, \beta, d_j) + \frac{1}{2} \cdot Pr(L_j \mid y_j = 1, \alpha, \beta, d_j) \quad (7)$$

Overall we obtain:

$$Pr[y_j = \lambda | L_j, \alpha, \beta, d_j] = \frac{Pr(L_j | y_j = \lambda, \alpha, \beta, d_j)}{Pr(L_j | y_j = 0, \alpha, \beta, d_j) + Pr(L_j | y_j = 1 \alpha, \beta, d_j)} \quad (8)$$

Let us consider one of these terms, and let us assume that every l_{ij} in L_j takes a value λ_p. We have

$$Pr(L_j \mid y_j = \lambda, \alpha, \beta, d_j) = \prod_{i=1}^{k} Pr(l_{ij} = \lambda_p \mid \alpha_i, \beta_i, d_j, y_j = \lambda) \quad (9)$$

If $\lambda_p = 0$ then $Pr(l_{ij} = \lambda_p \mid \alpha_i, \beta_i, d_j, y_j = 0)$ is the probability to classify correctly a 0 as 0, as defined in Eq. 4 denoted by $\delta_{ij} = \frac{1 + (1 - d_j)^{(1 - \beta_i)}}{2}$. Similarly, if $\lambda_p = 1$ then $Pr(l_{ij} = \lambda_p \mid \alpha_i, \beta_i, d_j, y_j = 1)$ is the probability to classify correctly a 1 as 1, expressed in Eq. 3 and denoted by $\gamma_{ij} = \frac{1 + (1 - d_j)^{(1 - \alpha_i)}}{2}$. Then the probability to classify $y_j = 1$ as $\lambda_p = 0$ is $(1 - \gamma_{ij})$ and the probability to classify $y_j = 1$ as $\lambda_p = 0$ is $(1 - \delta_{ij})$. We hence have $Pr(l_{ij} = \lambda_p \mid \alpha_i, \beta_i, d_j, y_j = 0) = (1 - \lambda_p) \cdot \delta_{ij} + \lambda_p \cdot (1 - \gamma_{ij})$. Similarly, we can write $Pr(l_{ij} = \lambda_p \mid \alpha_i, \beta_i, d_j, y_j = 1) = \lambda_p \cdot \gamma_{ij} + (1 - \lambda_p) \cdot (1 - \delta_{ij})$. So Eq. 8 rewrites as:

$$
\begin{aligned}
Pr[y_j = \lambda | L_j, \alpha, \beta, d_j] &= \frac{\prod_{i=1}^{k} Pr(l_{ij} = \lambda_p \mid y_j = \lambda_p), \alpha_i, \beta_i, d_j}{Pr(L_j \mid y_j = 0, \alpha, \beta, d_j) + Pr(L_j \mid y_j = 1, \alpha, \beta, d_j)} \\
&= \frac{\prod_{i=1}^{k} (1 - \lambda_p) \cdot [(1 - \lambda_p)\delta_{ij} + \lambda_p (1 - \gamma_{ij})] + \lambda_p \cdot [\lambda_p \cdot \gamma_{ij} + (1 - \lambda_p)(1 - \delta_{ij})]}{Pr(L_j \mid y_j = 0, \alpha, \beta, d_j) + Pr(L_j \mid y_j = 1, \alpha, \beta, d_j)} \\
&= \frac{\prod_{i=1}^{k} (1 - \lambda_p) \cdot [(1 - \lambda_p)\delta_{ij} + \lambda_p (1 - \gamma_{ij})] + \lambda_p \cdot [\lambda_p \cdot \gamma_{ij} + (1 - \lambda_p)(1 - \delta_{ij})]}{\prod_{i=1}^{k} (1 - \lambda_p)\delta_{ij} + \lambda_p (1 - \gamma_{ij}) + \prod_{i=1}^{k} \lambda_p \cdot \gamma_{ij} + (1 - \lambda_p)(1 - \delta_{ij})}
\end{aligned} \quad (10)
$$

In the E step, as every α_i, β_i, d_j is fixed, one can compute $\mathbb{E}[y_j | L_j, \alpha_i, \beta_i, d_j]$ and also choose as final value for y_j the value $\lambda \in \{0, 1\}$ such that $Pr[y_j = \lambda | L_j, \alpha_i, \beta_i, d_j] > Pr[y_j = (1 - \lambda_p) | L_j, \alpha_i, \beta_i, d_j]$. We can also estimate the likelihood for the values of variables $P(L \cup Y \mid \theta)$ for parameters $\theta = \{\alpha, \beta, d\}$, as $Pr(y_j = \lambda, L \mid \theta) = Pr(y_j = \lambda_p, L) \cdot Pr(L_j \mid y_j = \lambda_p, \theta) = Pr(y_j = \lambda_p) \cdot Pr(L_j \mid y_j = \lambda_p, \theta)$

M Step: With respect to the estimated posterior probabilities of Y computed during the E phase of the algorithm, we compute the parameters θ that maximize $Q(\theta, \theta^t)$. Let θ^t be the value of parameters computed at step t of the algorithm. We use the observed values of L, and the previous expectation for Y. We maximize $Q'(\theta, \theta^t) = \mathbb{E}[logPr(L, Y \mid \theta) \mid L, \theta^t]$ (we refer interested readers to [6]-Chap. 9 and [5] for explanations showing why this is equivalent to maximizing $Q(\theta, \theta^t)$). We can hence compute the next value as: $\theta^{t+1} = \arg\max_{\theta} Q'(\theta, \theta^t)$. Here in our context the values of θ are α_i, β_i, d_j. We maximize $Q'(\theta, \theta^t)$ using bounded optimization techniques, truncated Newton algorithm [11] provided by the standard SciPy[1] implementation. We iterate E and M steps, computing at each iteration t the posterior probability and the parameters θ^t that maximize $Q'(\theta, \theta^t)$. The algorithm converges, and stops when the improvement (difference between two successive joint log-likelihood values) is below a threshold, fixed in our case to $1e^{-7}$.

4   Cost Model

A drawback of many crowdsourcing approaches is that task distribution is static, i.e. tasks are distributed to a fixed number of workers, without considering their difficulty, nor checking if a consensus can be reached with fewer workers. Consider again the simple boolean tagging setting, but where each task realization are paid, and with a fixed total budget B_0 provided by the client. For simplicity, we assume that all workers receive 1 unit of credit for each realized task. Hence, to solve n boolean tagging tasks, one can hire only n/B_0 workers per task. In this section, we show a worker allocation algorithm that builds on collected answers and estimated difficulty to distribute tasks to worker at run time, and show its efficiency w.r.t. other approaches.

Our algorithm works in rounds. At each round, only a subset $T_{avl} \subseteq T$ of the initial tasks remain to be evaluated. We collect labels produced by workers for these tasks. We aggregate answers using the EM approach described in Sect. 3. We denote by y_j^q as the final aggregated answer for task j at round q, d_j^q is the current *difficulty* of task and α_i^q, β_i^q denotes the estimated *expertise* of a worker i at round q. We let $D^q = \{d_1^q \dots d_j^q\}$ denote the set of all difficulties estimated as round q. We fix a maximal step size $\tau \geq 1$, that is the maximal number of workers that can be hired during a round for a particular task. For every task $t_j \in T_{avl}$ with difficulty d_j^q at round q, we allocate $\mathbf{a}_j^q = \lceil (d_j^q / \max D^q) \times \tau \rceil$ workers for the next round. Once all answers for a task have been received, the EM aggregation can compute final label $y_j^q \in \{0, 1\}$, difficulty of task d_j^q, expertise of all workers $\alpha_1^q, \dots, \alpha_k^q, \beta_1^q, \dots, \beta_k^q$. Now, it remains to decide whether the confidence in answer y_j^q obtained at round q is sufficient (in which case, we do not allocate workers to this task in the next rounds). Let k_j^q be the number of answers obtained for task j at round q. The *confidence* $\hat{c}_j^q$ in a final label y_j^q is defined as follows:

---

[1] docs.scipy.org/doc/scipy/reference/generated/scipy.optimize.minimize.html.

$$\hat{c}_j^q(y_j^q = 1) = \frac{1}{k_j^q} \cdot \sum_{i=1}^{k_j^q} \left\{ l_{ij} \times \left(\frac{1+(1-d_j^q)^{(1-\alpha_i^q)}}{2}\right) + (1 - l_{ij}) \times \left(1 - \frac{1+(1-d_j^q)^{(1-\alpha_i^q)}}{2}\right) \right\} \quad (11)$$

$$\hat{c}_j^q(y_j^q = 0) = \frac{1}{k_j^q} \cdot \sum_{i=1}^{k_j^q} \left\{ (1 - l_{ij}) \times \left(\frac{1+(1-d_j^q)^{(1-\beta_i^q)}}{2}\right) + (l_{ij}) \times \left(1 - \frac{1+(1-d_j^q)^{(1-\beta_i^q)}}{2}\right) \right\} \quad (12)$$

Intuitively, each worker adds its probability of doing an error, which depends on the final label y_j^q estimated at round q and on his competences, i.e. on the probability to choose $l_{ij} = y_j^q$. Let us now show when to stop the rounds of our evaluation algorithm. We start with n tasks, and let T_{avl} denote the set of remaining tasks at round q. We define $r^q \in [0,1]$ as the ratio of task that are still considered at round q compared to the initial number of task, i.e. $r^q = \frac{|T_{avl}|}{n}$. We start with an initial budget B_0, and denote by B_c^q the total budget consumed at round q. We denote by $\mathcal{B}^q$ the fraction of budget consumed at that current instance, $\mathcal{B}^q = \frac{B_c^q}{B_0}$. We define the stopping threshold $Th^q \in [0.5, 1.0]$ as $Th^q = \frac{1+(1-\mathcal{B}^q)^{r^q}}{2}$.

The intuition behind this function is simple: when the number of remaining tasks decreases, one can afford a highest confidence threshold. Similarly, as the budget decreases, one shall derive a final answer for tasks faster, possibly with a poor confidence, as the remaining budget does not allow hiring many workers. Figure 1-(right) shows the different threshold based on the current estimate of the budget on the horizontal axis. Each line depicts the corresponding fraction of task available in the considered round. Observe that when r^q approaches 1, the threshold value falls rapidly, as large number of tasks remain without definite final answer, and have to be evaluated with the remaining budget. On the other hand, when there are less tasks (e.g. when $r^q = 0.10$), the threshold Th^q decreases slowly.

We can now define a crowdsourcing algorithm (CrowdInc) with a dynamic worker allocation strategy to optimize cost and accuracy. This strategy allocates workers depending on current confidence on final answers, and available resources. CrowdInc is decomposed in two phases, *Estimation* and *Convergence*.

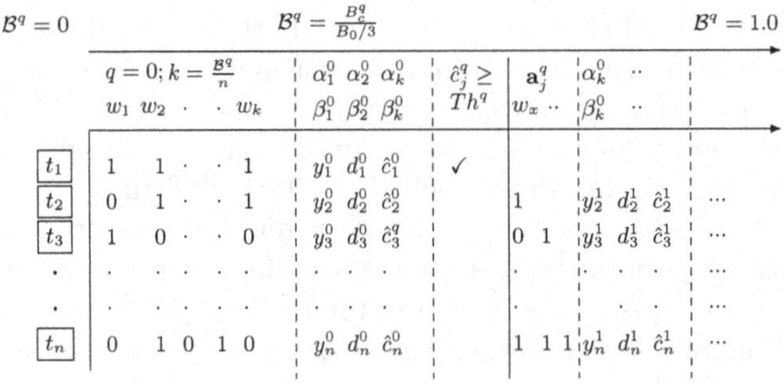

Fig. 2. A possible state for Algorithm 2

Estimation: As *difficulty* of tasks is not known a priori, the first challenge is to estimate it. To get an initial measure of difficulties, each task needs to be answered by a set of workers. Now, as each worker receives an incentive for a task, this preliminary evaluation has a cost, and finding an optimal number of workers for *difficulty* estimation is a fundamental issue. The initial budget gives some flexibility in the choice of an appropriate number of workers for preliminary evaluation of difficulty. Choosing a random number of workers per task does not seem a wise choice. We choose to devote a fraction of the initial budget to this estimation phase. We devote one third of the total budget ($B_0/3$) to the estimation phase. It leaves a sufficient budget ($2 \cdot B_0/3$) for the convergence phase. Experiments in the next Section show that this seems a sensible choice. After collection of answers for each task, we apply the EM based aggregation technique of Sect. 3 to estimate the *difficulty* of each task as well as the *expertise* of each worker. Considering this as an initial round $q = 0$, we let d_j^0 denote the initially estimated difficulty of each task j, and α_i^0, β_i^0 denote the expertise of each worker and y_j^0 denote the final aggregated answer. Note that if the difficulty of some tasks is available a priori and is provided by the client, we may skip the estimation step. However, in general clients do not possess such information and this initial step is crucial in estimation of parameters. After this initial estimation, one can already compute Th^0 and decide to stop evaluation of tasks with a sufficient confidence level.

Convergence: The *difficulty* of task d_j^q and the set of remaining tasks T_{avl} are used to start the convergence phase. Now as the difficulty of each task is estimated, we can use the estimated difficulty d_j^q to allocate the workers dynamically. The number of workers allocated at round $q > 0$ follows a difficulty aware *worker allocation* policy. At each round, we allocate $\mathbf{a}_j^q$ workers to remaining task t_j. This allocation policy guarantees that each remaining task is allocated at least one worker, at most τ workers, and that the more difficult tasks (i.e. have the more disagreement) are allocated more workers than easier tasks.

Algorithm 2 gives a full description of CrowdInc. We also show the information memorized at each step of the algorithms in Fig. 2. Consider a set of n tasks that have to be annotated with a boolean tag in $\{0, 1\}$. CrowdInc starts with the *Estimation* phase and allocates k workers for an initial evaluation round ($q = 0$). After collection of answers, and then at each round $q > 0$, we first apply EM based aggregation to estimate the difficulty d_j^q of each of task $t_j \in T_{avl}$, the confidence $\hat{c}_j^q$ in final aggregated answer y_j^q, and the expertise α_i^q, β_i^q of the workers. Then, we use the stopping threshold to decide whether we need more answers for each task. If $\hat{c}_j^q$ is greater than Th^q, the task t_j is removed from T_{avl}. This stopping criterion hence takes a decision based on the confidence in the final answers for a task and on the remaining budget. Consider, in the example of Fig. 2 that the aggregated answer for task t_1 has high confidence, and that $\hat{c}_j^q \geq Th^q$. Then, t_1 does not need further evaluation, and is removed from T_{avl}. Once solved tasks have been removed, we allocate $\mathbf{a}_j^q$ workers to each remaining task t_j in T_{avl} following our difficulty aware policy. Note that, each task gets a different number of workers based on task difficulty. The algorithm stops when

---

Algorithm 2: CrowdInc

Data: A set of tasks $T = \{t_1, \ldots, t_n\}$, a budget $= B_0$
Result: Final Answer: $Y = y_1, \ldots, y_n$, Difficulty: d_j, Expertise: α_i, β_i
1 **Initialization :** Set every d_j, α_i, β_i to a random value in $[0, 1]$.
2 $T_{avl} = T$; $q = 0$; $B = B - (B_0/3)$; $B_c = B_0/3$; $r = (B_0/3)/n$
3 //Initial Estimation:
4 Allocate r workers to each task in T_{avl} and get their answers
5 Estimate $d_j^q, \alpha_i^q, \beta_i^q, \hat{c}_j^q, 1 \leq j \leq n, 1 \leq i \leq B_0/3$ using EM aggregation
6 Compute the stopping threshold Th^q.
7 **for** $j = 1, \ldots, n$ **do**
8 $\quad$ | **if** $\hat{c}_j^q > Th^q$ **then** $T_{avl} = T \setminus \{j\}$;
9 **end**
10 //Convergence:
11 **while** $(B > 0)$ && $(T_{avl} \neq \emptyset)$ **do**
12 $\quad$ $q = q + 1$; $l = |T_{avl}|$
13 $\quad$ Allocate $\mathbf{a}_1^q, \ldots, \mathbf{a}_l^q$ workers to tasks $t_1, \ldots t_l$ based on difficulty.
14 $\quad$ Get the corresponding answers by all the newly allocated workers.
15 $\quad$ Estimate $d_j^q, \alpha_i^q, \beta_i^q, \hat{c}_j^q$ using aggregation model.
16 $\quad$ $B = B - \sum\limits_{i \in 1..|T_{avl}|} \mathbf{a}_i^q$
17 $\quad$ Compute the stopping threshold Th^q
18 $\quad$ **for** $j = 1, \ldots, n$ **do**
19 $\quad\quad$ | **if** $\hat{c}_j^q > Th^q$ **then** $T_{avl} = T_{avl} \setminus \{j\}$;
20 $\quad$ **end**
21 **end**

---

either all budget is exhausted or there is no additional task left. It returns the aggregated answers for all tasks.

5   Experiments

We evaluate the algorithm on three public available dataset, namely the product identification [17], duck identification [18] and Sentiment Analysis [12] benchmarks. We briefly detail each dataset and the corresponding tagging tasks. All tags appearing in the benchmarks were collected via Amazon Mechanical Turk.

In the **Product Identification** use case, workers were asked to decide whether a *product-name* and a *description* refer to the same product. The answer returned is *True* or *False*. There are 8315 samples and each of them was evaluated by 3 workers. The total number of unique workers is 176 and the total number of answers available is 24945. In the **Duck Identification** use case, workers had to decide if sample images contain a duck. The total number of tasks is 108 and each of task was allocated to 39 workers. The total number of unique worker is 39 and the total number of answers is 4212. In the **Sentiment Popularity** use case, workers had to annotate movie reviews as Positive or Negative opinions. The total number of tasks was 500. Each task was given to 20 unique workers and a total number of 143 workers were involved, resulting in a total number of 10000 answers. All these information are synthesized in Table 1.

Table 1. Datasets description.

Dataset	Number of tasks	Number of tasks with ground truth	Total Number of answers provided by crowd	Average number of answers for each task	Number of unique crowd workers
Product Identification	8315	8315	24945	3	176
Duck Identification	108	108	4212	39	39
Sentiment Popularity	500	500	10000	20	143

Table 2. Comparison of EM + aggregation (with Recall, specificity & task difficulty) w.r.t MV, D&S, GLAD, PMCRH, LFC, ZenCrowd.

Methods	Recall	Specificity	Balanced Accuracy	Methods	Recall	Specificity	Balanced Accuracy	Methods	Recall	Specificity	Balanced Accuracy
MV	0.56	0.91	0.73	MV	0.61	0.93	0.77	MV	0.93	0.94	0.4
D&S [3]	0.81	0.93	0.87	D&S [3]	0.65	0.97	0.81	D&S [3]	0.94	0.94	0.94
GLAD [19]	0.47	0.98	0.73	GLAD [19]	0.48	0.98	0.73	GLAD [19]	0.94	0.94	0.94
PMCRH [10]	0.58	0.95	0.76	PMCRH [10]	0.61	0.93	0.77	PMCRH [10]	0.93	0.95	0.94
LFC [14]	0.87	0.91	0.89	LFC [14]	0.64	0.97	0.81	LFC [14]	0.94	0.94	0.94
ZenCrowd [4]	0.39	0.98	0.68	ZenCrowd [4]	0.51	0.98	0.75	ZenCrowd [4]	0.94	0.94	0.94
EM + recall, specificity & difficulty	0.89	0.91	0.90	EM + recall, specificity & difficulty	0.77	0.90	0.83	EM + recall, specificity & difficulty	0.94	0.95	0.94

(a) Duck Identification   (b) Product Identification   (c) Sentiment Popularity

Evaluation of Aggregation: We first compared our aggregation technique to several methods: MV, D&S [3], GLAD [19], PMCRH [10], LFC [14], and Zen-Crowd [4]. We ran the experiment 30 times with different initial values for tasks difficulty and workers expertise. The standard deviation over all the iteration was less than 0.05%. Hence our aggregation is insensitive to initial prior values. We now compare *Recall*, *Specificity* and *Balanced Accuracy* of all methods. Balanced Accuracy is the average of recall and specificity. We can observe in Table 2 that our method outperforms other techniques in Duck Identification, Product Identification, and is comparable for Sentiment Popularity.

Evaluation of CrowdInc: The goal of the next experiment was to verify that the cost model proposed in CrowdInc achieves at least the same accuracy but with a smaller budget. We have used Duck identification and Sentiment popularity for this test. We did not consider the Product Identification benchmark: indeed, as shown in Table 1, the Product Identification associates only 3 answers to each task. This does not allow for a significant experiment with CrowdInc. We compared the performance of CrowdInc to other approaches in terms of cost and accuracy. The results are given in Fig. 3. Static(MV) denotes the traditional crowdsourcing platforms with majority voting as aggregation technique and Static(EM) shows more advanced aggregation technique with EM based

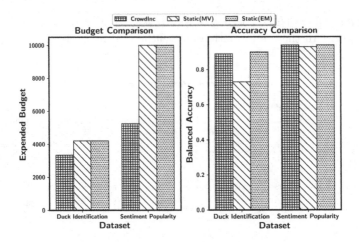

Fig. 3. Comparison of cost vs. Accuracy.

aggregation technique. Both algorithms allocate all the workers (and hence use all their budget) at the beginning of crowdsourcing process.

The following observation can be made from Fig. 3. First, CrowdInc achieves better accuracy than a static(MV) approach. This is not a real surprise, as MV already showed bad accuracy in Table 2. Then, CrowdInc achieves almost the same accuracy as a Static(EM) based approach in Duck identification, and the same accuracy in Sentiment Popularity. Last, CrowdInc uses a smaller budget than static approaches in all cases.

Table 3 shows the time (in seconds) needed by each algorithm to aggregate answers. Static(MV) is the fastest solution: it is not surprising, as the complexity is linear in the number of answers. We recall however that MV has the worst accuracy of all tested aggregation techniques. We have tested aggregation with EM when the number of workers is fixed a priori and is the same for all tasks (Static(EM)). CrowdInc uses EM, but on a dynamic sets of workers and tasks, stopping easiest tasks first. This results in a longer calculus, as EM is used several times on sets of answers of growing sizes. The accuracy of *static(EM)* and CrowdInc are almost the same. Aggregation with CrowdInc takes approximately 11% longer than *static(EM)* but for a smaller budget, as shown in the Fig. 3. To summarize the CrowdInc aggregation needs more time and a smaller budget to aggregate answers with a comparable accuracy. In general, clients using crowd-sourcing services can wait several days to see their task completed. Hence, when time is not a major concern CrowdInc can reduce the cost of crowdsourcing.

Table 3. Running time (in seconds) of CrowdInc, MV and Static EM.

Dataset/Methods	CrowdInc	Static(EM)	Static(MV)
Duck Identification	843.26	106.81	0.073
Sentiment Popularity	1323.35	137.79	0.102

6   Conclusion and Discussions

In this paper, we introduced an aggregation technique for crowdsourcing platforms. Aggregation is based on expectation maximization and jointly estimates the answers, the difficulty of tasks, and the expertise of workers. Using difficulty and expertise as latent variables improves the accuracy of aggregation in terms of recall and specificity. We also proposed CrowdInc an incremental labeling technique that optimizes the cost of answers collection. The algorithm implements a worker allocation policy that takes decisions from a dynamic threshold computed at each round, which helps achieving a trade off between cost and accuracy. We showed in experiments that our aggregation technique outperforms the existing state-of-the-art techniques. We also showed that our incremental crowdsourcing approach achieves the same accuracy as EM with static allocation of workers, better accuracy than majority voting, and in both cases at lower costs.

The ideas proposed in this paper can lead to several improvements that will be considered in future work. In the paper, we addressed binary tasks for simplicity, but the approach can be easily extended to tasks with a finite number m of answers. The difficulty of each task t_j remains a parameter d_j. Expertise is the ability to classify a task as m when its ground truth is m. An EM algorithm just has to consider probabilities of the form $Pr(L_{ij} = m|y_j = m)$ to derive hidden parameters and final labels for each task. An easy improvement is to consider incentives that depend on workers characteristics. This can be done with a slight adaptation of costs in the CrowdInc algorithm. Another possible improvement is to try to hire experts when the synthesized difficulty of a task is high, to avoid hiring numerous workers or increase the number of rounds. Another interesting topic to consider is the impact of answers introduced by a malevolent user on the final aggregated results.

Last, we think that the complexity of CrowdInc can be improved. The complexity of each E-step of the aggregation is linear in the number of answers. The M-step maximizes the log likelihood with an iterative process (truncated Newton algorithm). However, the E and M steps have to be repeated many times. The cost of this iteration can be seen in Table 3, where one clearly see the difference between a linear approach such as Majority Voting (third column), a single round of EM (second column), and CrowdInc. Using CrowdInc to reduce costs results in an increased duration to compute final answers. Indeed, the calculus performed at round i to compute hidden variables for a task t is lost at step $i+1$ if t is not stopped. An interesting idea is to consider how a part of computations can be reused from a round to the next one to speed up convergence.

References

1. Dai, P., Lin, C.H., Weld, D.S.: POMDP-based control of workflows for crowdsourcing. Artif. Intell. **202**, 52–85 (2013)
2. Daniel, F., Kucherbaev, P., Cappiello, C., Benatallah, B., Allahbakhsh, M.: Quality control in crowdsourcing: a survey of quality attributes, assessment techniques, and assurance actions. ACM Comput. Surv. **51**(1), 7 (2018)
3. Ph.A., Dawid, Skene, A.M.: Maximum likelihood estimation of observer error-rates using the EM algorithm. J. Royal Stat. Soc. Ser. C (Appl. Stat.) **28**(1), 20–28 (1979)
4. Demartini, G., Difallah, D.E., Cudré-Mauroux, Ph.: Zencrowd: leveraging probabilistic reasoning and crowdsourcing techniques for large-scale entity linking. In: Proceedings of WWW 2012, pp. 469–478. ACM (2012)
5. Dempster, A.P., Laird, N.M., Rubin, D.B.: Maximum likelihood from incomplete data via the EM algorithm. J. R. Stat. Soc. Ser. B (Methodol.) **39**(1), 1–22 (1977)
6. Flach, P.A.: Achine Learning - The Art and Science of Algorithms that MakeSense of Data. Cambridge University Press, Cambridge (2012)
7. Garcia-Molina, H., Joglekar, M., Marcus, A., Parameswaran, A., Verroios, V.: Challenges in data crowdsourcing. Trans. Knowl. Data Eng. **28**(4), 901–911 (2016)
8. Karger, D.R., Oh, S., Shah, D.: Iterative learning for reliable crowdsourcing systems. In: Proceedings of NIPS 2011, pp. 1953–1961 (2011)
9. Le, J., Edmonds, A., Hester, V., Biewald, L.: Ensuring quality in crowdsourced search relevance evaluation: the effects of training question distribution. In: SIGIR 2010 Workshop on Crowdsourcing for Search Evaluation, vol. 2126, pp. 22–32 (2010)
10. Li, Q., Li, Y., Gao, J., Zhao, B., Fan, W., Han, J.: Resolving conflicts in heterogeneous data by truth discovery and source reliability estimation. In: Proceedings of SIGMOD 2014, pp. 1187–1198. ACM (2014)
11. Nash, S.G.: Newton-type minimization via the Lanczos method. SIAM J. Num. Anal. **21**(4), 770–788 (1984)
12. Pang, B., Lee, L.: A sentimental education: sentiment analysis using subjectivity summarization based on minimum cuts. In: Proceedings of the 42nd annual meeting on Association for Computational Linguistics, pp. 271. Association for Computational Linguistics (2004)
13. Raykar, V., Agrawa, P.: Sequential crowdsourced labeling as an epsilon-greedy exploration in a Markov decision process. In: Artificial Intelligence and Statistics, pp. 832–840 (2014)
14. Raykar, V.C.: Learning from crowds. J. Mach. Learn. Res. **11**, 1297–1322 (2010)
15. Tran-Thanh, L., Venanzi, M., Rogers, A., Jennings, N.R.: Efficient budget allocation with accuracy guarantees for crowdsourcing classification tasks. In: Proceedings of AAMAS 2013, pp. 901–908. International Foundation for Autonomous Agents and Multiagent Systems (2013)
16. Venanzi, M., Guiver, J., Kazai, G., Kohli, P., Shokouhi, M.: Community-based Bayesian aggregation models for crowdsourcing. In: Proceedings of WWW 2014, pp. 155–164. ACM (2014)
17. Wang, J., Kraska, T., Franklin, M.J., Feng, J.: Crowder: crowdsourcing entity resolution. Proc. VLDB Endowment **5**(11), 1483–1494 (2012)
18. Welinder, P., Branson, S., Perona, P., Belongie, S.J.: The multidimensional wisdom of crowds. In: Proceedings of NIPS 2010, pp. 2424–2432 (2010)

19. Whitehill, J., Wu, T., Bergsma, J., Movellan, J.R., Ruvolo, P.L.: Whose vote should count more: optimal integration of labels from labelers of unknown expertise. In: Proceedings of NIPS 2009, pp. 2035–2043 (2009)
20. Zheng, Y., Li, G., Li, Y., Shan, C., Cheng, R.: Truth inference in crowdsourcing: is the problem solved? Proc. VLDB Endowment **10**(5), 541–552 (2017)

Web API Search: Discover Web API and Its Endpoint with Natural Language Queries

Lei Liu$^{(\boxtimes)}$, Mehdi Bahrami, Junhee Park, and Wei-Peng Chen

Fujitsu Laboratories of America, Inc.,
1240 E Arques Avenue, Sunnyvale, CA 94085, USA
{lliu,mbahrami,jpark,wchen}@fujitsu.com

Abstract. In recent years, Web Application Programming Interfaces (APIs) are becoming more and more popular with the development of the Internet industry and software engineering. Many companies provide public Web APIs for their services, and developers can greatly accelerate the development of new applications by relying on such APIs to execute complex tasks without implementing the corresponding functionalities themselves. The proliferation of web APIs, however, also introduces a challenge for developers to search and discover the desired API and its endpoint. This is a practical and crucial problem because according to ProgrammableWeb, there are more than 22,000 public Web APIs each of which may have tens or hundreds of endpoints. Therefore, it is difficult and time-consuming for developers to find the desired API and its endpoint to satisfy their development needs. In this paper, we present an intelligent system for Web API searches based on natural language queries by using a two-step transfer learning. To train the model, we collect a significant amount of sentences from crowdsourcing and utilize an ensemble deep learning model to predict the correct description sentences for an API and its endpoint. A training dataset is built by synthesizing the correct description sentences and then is used to train the two-step transfer learning model for Web API search. Extensive evaluation results show that the proposed methods and system can achieve high accuracy to search a Web API and its endpoint.

Keywords: Web APIs · Neural networks · Deep learning

1 Introduction

A Web API is an application programming interface exposed via the Web, commonly used as representational state transfer (RESTful) services through Hyper-Text Transfer Protocol (HTTP). As the Internet industry progresses, Web APIs become more concrete with emerging best practices and more popular for modern application development [1]. Web APIs provide an interface for easy software development through abstracting a variety of complex data and web services,

© Springer Nature Switzerland AG 2020
W.-S. Ku et al. (Eds.): ICWS 2020, LNCS 12406, pp. 96–113, 2020.
https://doi.org/10.1007/978-3-030-59618-7_7

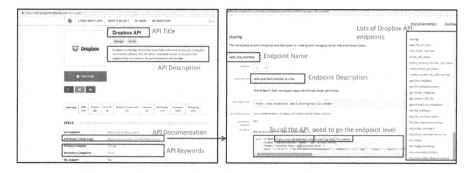

Fig. 1. Examples of ProgrammableWeb page of the Dropbox API (left) and documentations for Dropbox API (right).

which can greatly accelerate application development. Web APIs have also been widely used by technical companies due to their inherent flexibility. For example, Twitter offers public APIs to enable third parties to access and analyze historical tweets. Amazon provides free advertising APIs to developers as a way to promote their products. On the other hand, developers also benefit from the burgeoning API economy [2]. Developers can access various datasets and services via Web APIs and incorporate these resources into their development [3].

Due to these advantages, Web APIs have been widely developed in recent years. According to ProgrammableWeb[1], there are more than 22,000 public Web APIs available today, and this number is rapidly increasing. Moreover, an API has a number of endpoints, which specify the location of resources that developers need to access to carry out their functions. As an example shown in Fig. 1, the Dropbox API has 136 endpoints, and each endpoint has its own concrete function. In order to achieve the given function, an HTTP request has to be sent to the corresponding endpoint using a given HTTP method, as shown in Fig. 1.

The proliferation of Web APIs, however, makes it difficult for developers to search and discover a desired API and its endpoint. As aforementioned, the developer needs to know the endpoint in order to call an API. Therefore, the API level search is insufficient. In light of this, in this paper, we focus on building a Web API search system that can provide endpoint level search results based on a natural language query describing developers' needs. With the proposed dataset collection and generation methods and the two-step transfer learning model, the API search system can achieve high accuracy to search a Web API and its endpoint to satisfy developers' requirements.

2   Related Works and Our Contributions

API search or recommendation for developers has been extensively studied in the past. However, there are several key differences from our work:

---

[1] Available at: https://www.programmableweb.com.

Keyword-Based Search vs. Deep Learning-Based Search: Online Web API platforms such as ProgrammableWeb, Rapid API[2], and API Harmony[3] provides API search functions. However, strict keyword matching is used in these platforms to return a list of APIs given a user's query. Strict keyword search (syntactical representation) does not allow users to search semantically. On the other hand, deep learning-based method can enable semantic search, which means that users can search content for its meaning in addition to keywords, and maximize the chances the users will find the information they are looking for. For example, in the Fitbit API documentation, the word "glucose" is used. If the users search "blood sugar", the keyword-based search engine cannot return the Fitbit API. Thanks to the word embedding, the deep learning-based search engine, with semantic representation, is more advanced and intelligent to handle the scenario where exact keyword matching does not succeed. Supporting semantic search would help developers identify appropriate APIs in particular during the stage of application development when developers would not have clear ideas about which specific APIs to utilize or they would like to look for other similar APIs.

Programming-Language API Search vs. RESTful Web API Search: In the past decade, there are many works investigated search approaches for programming language APIs, for example, Java APIs or C++ APIs. [4] proposed RACK, a programming language API recommendation system that leverages code search queries from Stack Overflow[4] to recommend APIs based on a developer's query. The API text description has also been used for searching APIs. [5] proposed sourcerer API search to find Java API usage examples in large code repositories. [6] conducted a study by using a question and answering system to guide developers with unfamiliar APIs. [7] identified the problems of API search by utilizing Web search. Based on the observations, the authors presented a prototype search tool called Mica that augments standard Web search results to help programmers find the right API classes and methods given a description of the desired function as input, and help programmers find examples when they already know which methods to use. Compared to programming language API search, the Web API and its endpoint search is a new domain, and the major challenge is the lack of training datasets.

Web API-Level Search vs. Endpoint Level Search: There are several studies related to Web API search in recent years. [8] developed a language model based on the collected text such as the API description on the Web to support API queries. [10] used the API's multi-dimensional descriptions to enhance the search and ranking of Web APIs. The recent works in [11,12] investigated API level recommendation with a natural language query. As multiple APIs are frequently being used together for software development, many researchers have

---

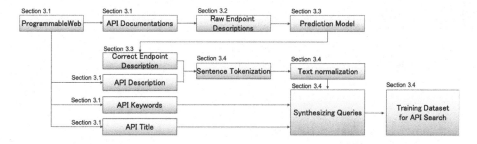

Fig. 2. Procedure for building the training dataset.

also focused on recommending API mashups, or recommending an API based on another API as the input, such as [13–17].

However, although there are many existing works regarding Web API search or recommendation, there is no study supporting API endpoint level discovery with natural language input. In addition, majority of the related works have been carried out using internal APIs or a limited number of APIs, not on a large number of public Web APIs. In practice, the recommendation with only API information is insufficient for developers because, in order to actually use the API in the application design, developers need endpoint level information. A web API may have tens or hundreds of endpoints. With only API level search results, it is still burdensome for developers to discover which endpoint should be used. To the best of our knowledge, this paper is the first work that proposes the endpoint level search on a large number of Web APIs whereas all the previous works only support API level search or recommendation. The novelty and contributions of this work are threefold:

- Currently, the key bottleneck for building a machine learning-based Web API search system is the lack of publicly available training datasets. Although the services like API Harmony provide structured API specification, where we can collect information such as API and its endpoint descriptions to build the training dataset, the majority of existing web APIs lack such structured API specifications. For example, API Harmony only supports 1,179 APIs, which is just a small percentage of all the Web APIs. In this paper, we propose a method to collect useful information directly from API documentation, and then build a training dataset, which can support more than 9,000 Web APIs for search purposes.
- As the information we collected, in particular, the endpoint descriptions from the API documentation may contain a lot of noise. We propose deep learning methods to predict correct API endpoint description sentences. The evaluation results show that decent accuracy can be achieved.
- We propose a two-step transfer learning method to support endpoint level Web API search, whereas all the previous works only support API level search. The evaluation results show that our proposed model can achieve high search accuracy.

3   Training Dataset Generation

In this section, the method for collecting and generating the training dataset is detailed. The entire procedure is shown in Fig. 2.

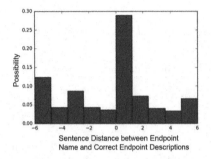

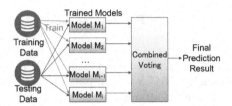

Fig. 3. Correct endpoint descriptions sentence span from endpoint name.

Fig. 4. Ensemble LSTM+ANN model to support endpoint description prediction.

3.1   Crowdsourcing from ProgrammableWeb

The information collection starts from ProgrammableWeb, which provides a directory service for more than 22,000 Web APIs. We can use web scraping to extract API data for each Web API. As an example, Fig. 1 shows the ProgrammableWeb page regarding the Dropbox API. From this page, we can collect the API title, API description and API keywords, which are essential for API level search. However, as we target on endpoint level search, we need to collect descriptions for each endpoint of the API, which are not available at ProgrammmableWeb. However, in programmableWeb page, we can find an URL for the documentation of the given API. In general, API documentation includes the list of endpoints and the description of each endpoint, as the example of the Dropbox API documentation shown in Fig. 1. Therefore, by further scraping the API documentation, we can collect data regarding endpoint descriptions.

However, by checking documentation of many Web APIs, we found it is a challenge to identify the correct endpoint descriptions. This is because there is no standard template for providers to write API documentations, and the quality of API documentation varies significantly. In some good quality and well-structured API documentation (such as the Dropbox API documentation in Fig. 1), the endpoint descriptions may directly follow the endpoint name, and there is an "description" tag for readers to easily to find the correct endpoint descriptions. On the other hand, in bad quality or poorly structured API documentation, the endpoint descriptions may have a certain distance from the endpoint name and without any tag, which are relatively hard to find.

We take the API Harmony data as an example to evaluate if there is any pattern regarding the distances between the endpoint name and its descriptions in API documentation. API Harmony is a catalog service of 1,179 web APIs, which provides structured data for these APIs according to the OpenAPI Specification (formerly Swagger Specification). From the structured data for a given API, we can retrieve its endpoints and the descriptions for each endpoint. We can assume these descriptions are ground-truth endpoint descriptions as they have been evaluated by users. We also collect the corresponding API documentation (i.e. a couple of HTML pages) for each given API from the websites of API providers using Web crawlers. After that, we check where the endpoint description is located in the corresponding documentation and how far it is from the endpoint name. The result is depicted in Fig. 3. It can be seen that only about 30% endpoint descriptions are the first sentence after the endpoint name. From this result, we can see that the correct endpoint descriptions can be 1 to 6 sentences before endpoint name or after endpoint name, so there is no particular pattern regarding the sentence distance between endpoint name and its correct endpoint descriptions in API documentation.

Based on the observation of the results in Fig. 3. We defined two terms: raw endpoint descriptions and correct endpoint descriptions. Raw endpoint descriptions are the sentences surrounding endpoint names in Web API documentation. For example, according to Fig. 3, we may define raw endpoint descriptions include 12 sentences (6 sentences before and 6 sentences after) for each endpoint name in API documentation. Among these raw endpoint descriptions, 1 or more sentences accurately describe the endpoint functions, and these sentences are referred to as correct endpoint descriptions. Such correct endpoint descriptions are essential in order to achieve accurate endpoint level search.

Due to the challenge of identifying the correct endpoint descriptions from raw endpoint descriptions, as well as the rapid growth of Web APIs, we need to design an automatic approach that can predict the correct endpoint descriptions from API documentation. To address this issue, in Sects. 3.2 and 3.3, we present how to collect raw endpoint descriptions from API documentation, and how to predict correct endpoint descriptions from raw endpoint descriptions, respectively.

3.2 Collection of Raw API Endpoint Descriptions from API Documentations

As aforementioned, raw endpoint descriptions are the surrounding sentences of each endpoint name in API documentation. An API may have a list of endpoints (E) and each endpoint is defined as E_i. For each API, we extract each of its endpoints (E_i) and raw description for this endpoint $(E_{i,D})$ from API documentation for a large number of APIs. We use a regular expression method, similar to [8], to extract the list of API endpoints (E). Regarding $E_{i,D}$, different API providers use different approaches to list endpoints and explain the description of endpoints. For example, some use a semi-structured information to list endpoints, some explain an endpoint and its description through a paragraph, and some use endpoint as header and explain the description. Our objective of information extraction is collecting all possible $E_{i,D}$ for each endpoint E_i in each HTML page, where $i \in [1, ..l]$ and l represents the total number of end-points in one HTML page. We extract information from semi-structured pages by processing HTML table tags and the table headers. To this end, we define a placeholder $\mathcal{P}_i$ that contains both E_i and $E_{i,D}$. $\mathcal{P}_i$ represents a section of the HTML page which appears between two same HTML headers ($[h_1, h_2, ..., h_6]$), and E_i is located in the section. Therefore, a raw endpoint description is the text around API endpoint as $E_{i_D} = [S_M, S_N]$ which denotes M sentences before and N sentences after appearing E_i inside $\mathcal{P}_i$. Algorithm 1 explains the detailed information of extracting endpoint description for the given endpoint list. By using the proposed method and setting $M = 6$ and $N = 6$, we collected $2,822,997$ web-pages with the size of $208.6\,$GB for more than $20,000$ public APIs. Such huge raw data contains a lot of noises. Therefore, in Sect. 3.3, we propose a deep learning method to predict and filter the correct endpoint descriptions from the raw descriptions.

Algorithm 1. Raw Endpoint Description Extraction

1: **procedure** GET_RAW_DESC(HTML,E,M,N)
    ▷ Extracts raw endpoint description $E_{i,D} \; \forall \; i \; \in \; E$
2:    $root = html.root$
3:    **for** E_i in E **do**
4:        $Tag = Find\;Tag\;of\;E_i\;in\;html$
5:        **if** Tag!=Null **then**
6:            **while** Tag!=root **do**
7:                **for** h_i in $[h_1, h_2, ..., h_6]$ **do**
8:                    **if** $h_i == Tag.name$ **then**
9:                        $\mathcal{P}_i = Tag$
10:                        $Break$ **else**
11:                            $Tag = Tag.parent$
12:        $sents = sent\_token(\mathcal{P}_i.content)$
13:        $pos = find(E_i)\;in\;sents$
14:        $raw\_sents = sents[pos - M : pos + N]$
15:        return raw_sents

3.3   Prediction of Correct Endpoint Descriptions

Deep Learning Models: We propose a deep learning method to predict the correct endpoint description sentences from raw endpoint descriptions for each endpoint. To train the deep learning model, we generated a training dataset based on API Harmony. For each endpoint of the APIs in API Harmony, a correct description is presented, which can be considered as ground-truth. For each API in API Harmony, we directly collect its documentation. After that, for each of the endpoint that this API has, we extract M sentences before and N sentences after the given endpoint name from the documentations, by using the algorithm presented in Sect. 3.2. Next, we compare the similarity for each of these $M + N$ sentences with the ground-truth endpoint description in API Harmony using spaCy[5], which calculates the similarity score by comparing word vectors. The sentence with the highest similarity can be considered as the correct endpoint description (i.e. ground-truth selected by API Harmony). The remaining $N + M - 1$ sentences which are not selected by API Harmony as endpoint descriptions can be treated as incorrect endpoint descriptions.

If the ground-truth endpoint description in API Harmony for a given endpoint contains K sentences where $K > 1$, in this case, "K-grams" of M sentences before and N sentences after need to be generated. For example, if $K = 2$ which means ground-truth endpoint description contains two sentences (GT_{S1}, GT_{S2}), we need to collect "2-grams" sentence pairs (T_i, T_j) from API documentation, such as (before 3rd, before 2nd), (before 2nd, before 1st), (after 1st, after 2nd), (after 2nd, after 3rd) where "before 3rd" means the 3rd sentence before endpoint name. After that, the average similarity score is computed according to the following equation:

$$Sim_{score} = (Sim(GT_{S1}, T_i) + Sim(GT_{S1}, T_j) \\ + Sim(GT_{S2}, T_i) + Sim(GT_{S2}, T_j))/4 \tag{1}$$

where Sim represents the similarity score between two given inputs. Similarly, the "K-gram" with the highest similarity can be considered as the correct endpoint descriptions (i.e. selected by API Harmony). The remaining "K-grams" which are not selected by API Harmony can be treated as incorrect endpoint descriptions.

For each correct or incorrect endpoint descriptions (with a label 1 or 0), we compute the following features to be used for the deep learning models:

- Endpoint Vector: Vector representation of endpoint names.
- Description Vector: Vector representation of correct or incorrect description sentences.
- HTTP Verb: HTTP method verbs (such as GET, POST, PUT, PATCH, DELETE, OPTIONS, HEAD) presented in the given sentence. If no such verb in the sentence, mark it as NONE. Those keywords are encoded by one-hot labels.
- Cosine Similarity: Cosine similarity between Endpoint and Description Vectors.
- spaCy Similarity: The average similarity score between the endpoint and description text calculated by SpaCy.
- HTML Section: Check if the given sentence and the endpoint name are in the same HTML section by checking the HTML tag. If yes, mark the sentence as "1", otherwise "0".
- Description Tag: Check if there is any HTML header tag with name "Description" or "description" in the given HTML section. If yes, mark the sentence as "1", otherwise "0".

---

[5] Available at: https://spacy.io/.

- Number of Tokens: Number of words including special characters in the description.
- Endpoint Count: Number of times that the endpoint name is found in the given sentence.

Note that the features for the description classification model are selected by observing ground-truth examples. For example, we observed that in many cases, the endpoint name and its description are similar, then we utilized spaCy similarity as one of the features. Extensive convincing examples can be found: in Gmail API, the endpoint "/userId/drafts" has the description "lists the drafts in the user's mailbox"; in Spotify API, the endpoint "/albums" has the description "get several albums" etc. Other features are also based on such observations.

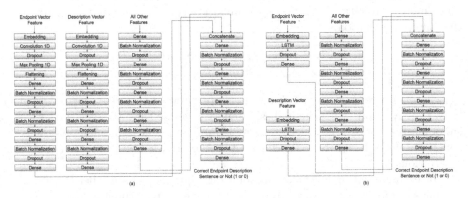

Fig. 5. Proposed deep learning models to predict correct endpoint descriptions: (a) CNN+ANN; (b) LSTM+ANN.

In many cases, the endpoint name is concatenated by multiple words, such as "/UserID". To compute word vectors for it, we firstly split such endpoint names into individual words, such as "User" and "ID", and then get the corresponding word embedding. Such splitting is achieved by building a word frequency list according to all the words that appeared in API and endpoint description sentences. We assume the words with higher frequency will have a lower cost for the splitting. By using dynamic programming, we can split such endpoint names with the target of minimizing the overall splitting cost. The GloVe [18] pre-trained word vectors with 6B tokens, 400K vocabulary, and 300-dimensional vectors are used for word embedding. If the endpoint and description have multiple words, we get their vectors by averaging the embedding of all the words.

The deep learning models predict whether a given sentence is a correct endpoint description or not. Fig. 5(a) presents our designed convolutional neural network (CNN)+ANN model and Fig. 5(b) shows a long short-term memory (LSTM)+ANN model. Here, ANN refers to an artificial neural network with Dense, Batch Normalization and Dropout layers. The inputs to the models are the features aforementioned, and the output is a binary indication (0 or 1) representing whether the sentence is a correct endpoint description or not for the given endpoint. In CNN+ANN model, the endpoint vectors and description vectors are sent to CNNs, and all the other features are sent to an ANN. The outputs of the CNNs and the ANN are merged and then sent to another ANN with multiple Dense, Dropout and Batch Normalization layers. The overall architecture for the LSTM+ANN model is similar to the CNN+ANN model. The only difference is that endpoint vector and description vector features are sent to LSTM networks, rather than CNNs, as shown in Fig. 5(a) and Fig. 5(b) respectively.

Performance Evaluation of Models: By using the above method, we extract the training dataset from API documentation, as summarized in Table 1. Note that correct and incorrect endpoint description sentences are imbalanced because only a small percentage of sentences are correct endpoint description sentences (selected by API Harmony), whereas all the rest sentences are incorrect description sentences. Therefore, we collect more incorrect endpoint description sentences compared with correct endpoint description sentences.

Table 1. Collected training dataset.

Training Dataset	# of Records
Correct endpoint description sentences	5,464
Incorrect endpoint description sentences	33,757

Table 2. Testing results for the deep learning models in Fig. 5 and traditional machine learning models.

Models	Testing accuracy
Decision Tree [19]	76.64%
Random Forest [20]	79.92%
CNN+ANN (Fig. 5(a))	90.31%
LSTM+ANN (Fig. 5(b))	98.13%

Since the training dataset for correct and incorrect endpoint description sentences is imbalanced, we firstly randomly select 5,588 sentences out of the 33,757 incorrect endpoint description sentences, and together with the 5,464 correct endpoint description sentences, we train the deep learning models depicted in Fig. 5. We use 65%, 20%, and 15% of the dataset for training, validation, and testing respectively. The testing result is shown in Table 2, which shows that both CNN+ANN and LSTM+ANN models can achieve more than 90% testing accuracy, and the LSTM+ANN model outperforms the CNN+ANN model. For comparison purposes, we also evaluate the performance of two traditional learning models: Decision Tree and Random Forest. Decision Tree is a flowchart graph or diagram that helps explore all of the decision alternatives and their possible outcomes [19]. Random Forest is an ensemble learning method for classification, regression, and other tasks, that operates by constructing a multitude of decision trees at training time and outputting the class based on voting [20]. The testing result in Table 2 shows that the proposed deep learning models greatly outperform the traditional learning models such as Decision Tree and Random Forest.

Blind Testing and Model Improvement: In the above testing, the training and testing datasets are all retrieved from API documentation related to the APIs included in API Harmony. However, API Harmony only covers a small percentage of Web APIs, and most of these APIs are made by big providers which are likely to have high-quality documentations. However, as we are targeting a wider coverage of Web APIs in the recommendation system, it is essential to evaluate the model performance over a large API documentation corpus, in particular for those not covered by API Harmony.

To conduct this blind testing, we manually label 632 sentences in documentations of APIs that are not covered by API Harmony. We compute all the features of these 632 sentences and send them as input to our trained LSTM+ANN model aforementioned. The results are summarized in Table 3. From the results, we can see that with only one trained model, the blind testing performance is not good as the model cannot distinguish the incorrect endpoint descriptions well. The reason is that when we train the model, we use the random under-sampling method in order to have a balanced training dataset between correct and incorrect description sentences. However, this method may discard potentially useful information which could be important for training the model. The samples chosen by random under-sampling may be biased samples, and thus, they may not be an accurate representation and provide sufficient coverage for incorrect descriptions, thereby, causing inaccurate results. To improve the model to cover a wider range of APIs, we applied an ensemble approach, as shown in Fig. 4.

Table 3. Blind testing results of ensemble method with multiple models.

# of Models	Accuracy	Recall	Precision
1 Model	31.80%	96.05%	14.57%
3 Models	78.80%	69.74%	32.32%
5 Models	80.70%	76.32%	35.80%
7 Models	84.97%	82.99%	43.45%

Table 4. Summary of training dataset.

Dataset	Number of APIs	Number of endpoints	Number of queries
API Harmony	1,127	9,004	232,296
Popular API List	1,603	12,659	447,904
Full API List	9,040	49,083	1,155,821

In Fig. 4, each model M_i is one trained model mentioned above. Here, the LSTM+ANN model is used as it outperforms CNN+ANN. Each M_i is trained with correct endpoint description sentences and different incorrect endpoint description sentences. This is achievable because we have much more incorrect endpoint descriptions sentences than the correct ones. In this case, each

M_i makes different decisions based on the learned features. They can predict independently and vote to jointly decide whether the input sentence is correct endpoint descriptions or not.

Table 3 shows the performance of the ensemble approach. It can be seen that the ensemble approach can improve the overall performance in terms of accuracy and precision, compared with only one model. Moreover, the ensemble approach with 7 models outperforms others, which will be used in the rest of this paper. The only issue is that some incorrect endpoint descriptions are wrongly predicted as correct endpoint descriptions, which result in more false-positive predictions and will introduce some noise to the training dataset of the API search model.

3.4   Synthesizing Queries for Training Dataset

In the previous steps, we have collected the data regarding API titles, API keywords, API descriptions, and correct endpoint descriptions. The API descriptions and correct endpoint descriptions may contain many sentences. Therefore, we firstly conduct sentence tokenization, and in turn, for each of the tokenized sentence, text normalization is carried out, including conducting word-stemming lemmatization and removing stop words, symbols, special characters, HTML tags, unnecessary spaces, and very short description sentence with only 1 word. After that, these processed sentences are used to build the training dataset.

We consider there are 4 major types of queries when a developer wants to search a Web API:

– Question type queries: developers may enter a question to search a Web API, for example, a question type query might be "which API can get glucose?"
– Command type queries: instead of asking a question, developers may directly enter a command type query to search an API, such as "get weather information."
– Keyword type queries: in many cases, developers may just input a couple of keywords to search an API. One example query is "fitness, health, wearable."
– API title-based queries: in some cases, developers may already have an idea regarding what API to use. Developers may just need to search an endpoint for this given API. One example of such a query is "post photos to Instagram." In this case, the search engine should return the endpoint of the Instagram API, rather than the endpoint of other similar APIs.

We define rule-based methods to synthesize training queries based on part-of-speech (POS) tagging and dependency parsing (also known as syntactic parsing). POS tagging is the process of marking up a word in a text as corresponding to a particular part of speech, based on both its definition and its context. Dependency parsing is the task of recognizing a sentence and assigning a syntactic structure to it. Figure 6(a) shows an example sentence with its POS tagging and dependency parsing results, which can be generated by many NLP tools such are

Spacy, NLTK, CoreNLP, etc. In this work, we use SpaCy, and the annotation of the POS tagging[6] and dependencies[7] can be found in SpaCy documentations.

Considering the fact that most of the sentences in API descriptions and endpoint descriptions are long, whereas in real practice, developers are unlikely to enter a very long query to a search engine, therefore we use POS tagging and dependency parsing to synthesize simplified question-type and command-type queries. We defined several rules, and if the description sentence satisfies a rule, simplified question-type and command-type queries are generated. For example, a rule is defined as

```
for subject in sentence:
    If subject.dep == nsubj and subject.head.pos == VERB:
        # Simplified question-type query:
        Q_query = \Which endpoint" + VERB + dobj NOUN phrase
        # Simplified command-type query:
        C_query = VERB + dobj NOUN phrase
```

Such a rule is feasible because the syntactic relations form a tree, every word has exactly one head. We can, therefore, iterate over the arcs in the dependency tree by iterating over the words in the sentence. If the original endpoint description sentence is "this endpoint gets a music playlist according to an artist ID," by using the rule, we can generate a simplified question-type query "which endpoint get a music playlist?", and a simplified command-type query "get a music playlist". The training dataset includes the API and endpoint description sentence, as well as the simplified question-type query and simplified command-type query. If an API or endpoint description sentence cannot be applied to any of the pre-defined rules, no simplified question-type query and simplified command-type query can be generated. In this case, only the API or endpoint description sentences will be included in the training dataset.

The keyword-based queries are generated from the API keywords that we collected from ProgrammableWeb. For example, the Spotify API has two category keywords "music" and "data mining" on ProgrammableWeb. So the keyword query can be "music, data mining". The keyword-based query can also be generated by concatenating the noun phrases of an API or endpoint description sentence. Given the same example, "this endpoint gets a music playlist according to an artist ID," the corresponding keyword-based query is "this endpoint, a music playlist, an artist ID."

The API title-based queries can be generated by using the API title collected from ProgrammableWeb. In addition, to emulate an API title-based query, we also attach the API title to the end of the short question-type queries and command-type queries. For the same example, the API title-based queries are "which endpoint get a music playlist with Spotify?" and "get a music playlist with Spotify."

---

[6] https://spacy.io/api/annotation#pos-tagging.

[7] https://spacy.io/api/annotation#dependency-parsing.

By using the proposed methods, we can build a training dataset for API/ endpoint search. The dataset has 3 columns, which are the query, its corresponding API and endpoint respectively. Note that, we cannot judge which endpoint should be used for the synthesized queries related to API description, API title, and ProgrammableWeb keywords. In this case, the endpoint field in the training dataset regarding these queries is marked as "N/A".

4   Web API Search: Deep Learning Model and Performance Evaluation

4.1   Deep Learning Model

The target of the API search engine is to search both API and its endpoint based on a natural language query. In this paper, we propose a two-step transfer learning method. The proposed model was designed by a performance comparison of multiple different architectures, and we picked up the model with the best performance. Figures 6(b) and (c) show the first step, which is to predict API and endpoint separately, and Fig. 6(d) shows the second step, which predicts API and its endpoints jointly, by reusing the models trained in the first step. The recurrent neural network model has also been widely used for text modeling. Especially, LSTM is gaining popularity as it specifically addresses the issue of learning long-term dependencies [21]. Therefore, we implement the recommendation model based on LSTM networks. As shown in Fig. 6(b) and 6(c), LSTM models in the first step include four layers: an input layer to instantiate a tensor and define the input shape, an embedding layer to convert tokens into

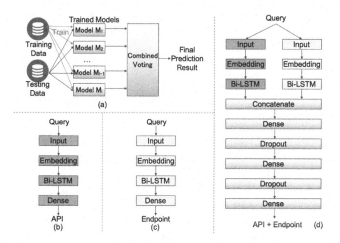

Fig. 6. (a) Ensemble LSTM+ANN model to support endpoint description prediction; (b) Bidirectional LSTM model to predict API (first step); (c) Bidirectional LSTM model to predict endpoint (first step); (d) Final model to predict both API and its endpoint (second step).

vectors, a bidirectional LSTM layer to access the long-range context in both input directions, and a dense layer with softmax activation function to linearize the output into a prediction result.

The training data that we used to train Fig. 6(b) and Fig. 6(c) are different. The model in Fig. 6(b) is trained by query and its corresponding API data, while the model in Fig. 6(c) is trained by query and its corresponding endpoint data. We fine-tune parameters in different models in order to make the training process consistent across different models. We train all models with a batch size of 512 examples. The maximum sentence length is 25 tokens. The GloVe [18] pre-trained word vectors with 6B tokens, 400K vocabulary, and 300-dimensional vectors are used for word embedding. A sentence is represented by averaging the embedding of all the words. We choose 500 hidden units for bi-directional LSTM.

After the two models in Fig. 6(b) and Fig. 6(c) are trained and well-fitted, we re-use these models for the second step to build the final model to predict both API and its endpoint simultaneously. The network architecture is shown in Fig. 6(d). The last dense layer of models in Fig. 6(b) and Fig. 6(c) is removed, and the rest layers are re-used in the final model. The parameters and weights in the corresponding layers are frozen, which means that these layers are not trainable when we train the final model in step 2. A concatenate layer is deployed to merge the output from two bidirectional LSTM layers. In turn, the output of the concatenate layer is sent to a neural network with dense layers and dropout layers. The final prediction is given by the last dense layer with softmax activation function. The loss function is categorical cross-entropy loss. The query and API/endpoint combination in the training dataset is used to train the final model. We set the dropout rate to 0.4 and used early stopping to avoid overfitting. The size of the output layer for models in Fig. 6(b), Fig. 6(c), and Fig. 6(d) equals to the number of APIs, number of endpoints, and number of API/endpoint pairs, respectively.

4.2   Performance Evaluation

To evaluate the performance of the API search, we test the model performance for the following dataset:

- API Harmony: As aforementioned, we consider API Harmony as the ground-truth dataset for API endpoint descriptions. Therefore, the testing results of the API search model for API Harmony can validate the overall efficiency of the proposed model when the training dataset is accurate.
- Full API list: The full API list is a comprehensive dataset that we collected in Sect. 3. The full API list covers 9,040 APIs. This number is smaller than the number of APIs in ProgrammableWeb because some of the documentation links in ProgrammableWeb cannot be accessed or the documentations are lacking endpoint information.
- Popular API list: We use the full API list and extract metadata about APIs from ProgrammableWeb and GitHub to rank the APIs in the full API list

based on their popularity. The popularity rank is computed based on the following extracted subjects where each one contains a number for a given API: (1) number of SDKs, (2) number of articles, (3) number of changelogs, (4) number of sample source codes, (5) number of "how to" articles, (6) number of libraries, (7) number of developers, (8) number followers, and (9) number of Github projects using this API. Items (1)–(8) are collected directly from ProgrammableWeb. Item (9) is collected from GitHub projects by searching API's host address and base path via GitHub APIs. The numbers collected in (1)-(9) are normalized and considered with the same weight for ranking API popularity. Based on the final ranking, we select the top 1,000 APIs for this dataset. If the providers of the top 1,000 APIs have other APIs which are not ranked in the top 1,000, we also add those APIs into this dataset. By doing so, the popular API list covers 1,603 APIs, which can be considered as the most popular Web APIs.

Performance Evaluation: The summary of the training dataset of the API Harmony, popular API list, and full API list is shown in Table 4. The training dataset is split into 80% for training and 20% for testing. The testing accuracy for the API search model is shown in Table 5. In this table, the top 1 accuracy shows the possibility that the correct API/endpoint is ranked as the first search result. Similarly, the top 10 accuracy represents the possibility that the correct API/endpoint is ranked as one of the first 10 search results. All the APIs/endpoints in search results are ranked by the possibility score given by the softmax function. This evaluation result shows that the proposed method can achieve very good accuracy for endpoint level search.

We compare the performance of the proposed 2 step transfer learning with the models that use traditional LSTM [21] or bi-LSTM [22] to recommend both API and its endpoint using the API Harmony dataset. The result is shown in Table 6, which validates that the proposed 2 step transfer learning model outperforms previous LSTM or bi-LSTM in terms of endpoint search accuracy.

Table 5. API/endpoint search accuracy.

Input Dataset	Top 1 accuracy	Top 10 accuracy
API Harmony	91.13%	97.42%
Popular API List	82.72%	93.81%
Full API List	78.85%	89.69%

Table 6. Comparison of the proposed 2 step transfer learning model with the LSTM and bi-LSTM for endpoint level search by using API Harmony dataset.

Model	Top 1 accuracy	Top 10 accuracy
LSTM	72.15%	80.27%
Bi-LSTM	74.48%	82.98%
2 step transfer learning	91.13%	97.42%

5   Conclusions

In this paper, we propose novel approaches to support an end-to-end procedure to build a Web API search system on a large number of public APIs. To the best of our knowledge, it is the first work that provides API endpoint level searches with a large API coverage (over 9,000 APIs) and high search accuracy. Our future work is to open the system to the public and collect users' query and their feedback. It is worth noting that the problem/application of Web API search is very practical for both academia and industry. Considering the fact that the state-of-the-art works only has a small API coverage (e.g. 1,179 APIs in API Harmony), constructing an API search system with 9,040 APIs and 49,083 endpoints is a significant improvement to this application. As Web APIs are rapidly growing and becoming more and more important for future software engineering, we hope the proposed application and its associated methods would be beneficial for the whole community.

References

1. Richardson, L., Ruby, S.: RESTful web services. O'Reilly Media Inc., Reading (2008)
2. Tan, W., Fan, Y., Ghoneim, A., et al.: From the service-oriented architecture to the Web API economy. IEEE Internet Comput. **20**(4), 64–68 (2016)
3. Verborgh, R., Dumontier, M.: A Web API ecosystem through feature-based reuse. IEEE Internet Comput. **22**(3), 29–37 (2018)
4. Rahman, M.M., Roy, C., Lo, D.: Rack: automatic API recommendation using crowdsourced knowledge. In: IEEE 23rd International Conference on Software Analysis, Evolution, and Reengineering (SANER), pp. 349–359. IEEE (2016)
5. Bajracharya, S., Ossher, J., Lopes, C.: Searching API usage examples in code repositories with sourcerer API search. In: ICSE Workshop on Search-driven Development: Users, Infrastructure, Tools and Evaluation, pp. 5–8. ACM (2010)
6. Duala-Ekoko, E., Robillard, M.: Asking and answering questions about unfamiliar APIs: An exploratory study. In: 34th International Conference on Software Engineering (ICSE), pp. 266–276. IEEE (2012)
7. Stylos, J., et al.: MICA: a web-search tool for finding API components and examples. In: Visual Languages and Human-Centric Computing, pp. 195–202. IEEE (2006)
8. Bahrami, M., et al.: API learning: applying machine learning to manage the rise of API economy. In: Proceedings of the Web Conference, pp. 151–154. ACM (2018)
9. Gu, X., et al.: Deep API learning. In: Proceedings of the International Symposium on Foundations of Software Engineering, pp. 631–642. ACM (2016)
10. Bianchini, Devis., De Antonellis, Valeria, Melchiori, Michele: A multi-perspective framework for web API search in enterprise mashup design. In: Salinesi, Camille, Norrie, Moira C., Pastor, Óscar (eds.) CAiSE 2013. LNCS, vol. 7908, pp. 353–368. Springer, Heidelberg (2013). https://doi.org/10.1007/978-3-642-38709-8_23
11. Su, Y., et al.: Building natural language interfaces to web APIs. In: Proceedings of Conference on Information and Knowledge Management, pp. 177–186. ACM (2017)

12. Lin, C., Kalia, A., Xiao, J., et al.: NL2API: a framework for bootstrapping service recommendation using natural language queries. In: IEEE International Conference on Web Services (ICWS), pp. 235–242. IEEE (2018)
13. Torres, R., Tapia, B.: Improving web API discovery by leveraging social information. In: IEEE International Conference on Web Services, pp. 744–745. IEEE (2011)
14. Cao, B., et al.: Mashup service recommendation based on user interest and social network. In: International Conference on Web Services, pp. 99–106. IEEE (2013)
15. Li, C., et al.: A novel approach for API recommendation in mashup development. In: International Conference on Web Services, pp. 289–296. IEEE (2014)
16. Gao, W., et al.: Manifold-learning based API recommendation for mashup creation. In: International Conference on Web Services, pp. 432–439. IEEE (2015)
17. Yang, Y., Liu, P., Ding, L., et al.: ServeNet: a deep neural network for web service classification. arXiv preprint arXiv:1806.05437 (2018)
18. Pennington, J., Socher, R., Manning, C.: Glove: global vectors for word representation. In: Proceedings of the Conference on Empirical Methods in Natural Language Processing (EMNLP), pp. 1532–1543. ACM (2014)
19. Apté, C., Weiss, S.: Data mining with decision trees and decision rules. Fut. Gen. Comput. Syst. 13(2–3), 197–210 (1997)
20. Shi, T., Horvath, S.: Unsupervised learning with random forest predictors. J. Comput. Graph. Stat. 15(1), 118–138 (2006)
21. Jozefowicz, R., et al.: An empirical exploration of recurrent network architectures. In: International Conference on Machine Learning, pp. 2342–2350 (2015)
22. Schuster, M., Paliwal, K.: Bidirectional recurrent neural networks. IEEE Trans. Signal Process. 45(11), 2673–2681 (1997)

Web Service API Anti-patterns Detection as a Multi-label Learning Problem

Islem Saidani[1], Ali Ouni[1(✉)], and Mohamed Wiem Mkaouer[2]

[1] Ecode de technologie superieure, University of Quebec, Montreal, QC, Canada
islem.saidani@ens.etsmtl.ca, ali.ouni@etsmtl.ca
[2] Rechester Institute of Technology, Rochester, NY, USA
mwmvse@rit.edu

Abstract. Anti-patterns are symptoms of poor design and implementation solutions applied by developers during the development of their software systems. Recent studies have identified a variety of Web service anti-patterns and defined them as sub-optimal solutions that result from bad design choices, time pressure, or lack of developers experience. The existence of anti-patterns often leads to software systems that are hard to understand, reuse, and discover in practice. Indeed, it has been shown that service designers and developers tend to pay little attention to their service interfaces design. Web service antipatterns detection is a non-trivial and error-prone task as different anti-pattern types typically have interleaving symptoms that can be subjectively interpreted and hence detected in different ways. In this paper, we introduce an automated approach that learns from a set of interleaving Web service design symptoms that characterize the existence of anti-pattern instances in a service-based system. We build a multi-label learning model to detect 8 common types of Web service anti-patterns. We use the ensemble classifier chain (ECC) model that transforms multi-label problems into several single-label problems which are solved using genetic programming (GP) to find the optimal detection rules for each anti-pattern type. To evaluate the performance of our approach, we conducted an empirical study on a benchmark of 815 Web services. The statistical tests of our results show that our approach can detect the eight Web service antipattern types with an average F-measure of 93% achieving a better performance compared to different state-of-the-art techniques. Furthermore, we found that the most influential factors that best characterize Web service anti-patterns include the number of declared operations, the number of port types, and the number of simple and complex types in service interfaces.

Keywords: Web service design · Service interface · Service anti-patterns · Genetic programming · Ensemble classifier chain

1 Introduction

Web services have become a popular technology for deploying scale-out application logic and are used in both open source and industry software projects such

© Springer Nature Switzerland AG 2020
W.-S. Ku et al. (Eds.): ICWS 2020, LNCS 12406, pp. 114–132, 2020.
https://doi.org/10.1007/978-3-030-59618-7_8

as Amazon, Yahoo, Fedex, Netflix, and Google. An advantage of using Web services and Service-Based Systems (SBS) is their loose coupling, which leads to agile and rapid evolution, and continuous re-deployment. Typically, SBSs use of a collection Web services that communicate by messages through declared operations in the services interfaces (API).

Being the most used implementation of the Service Oriented Architecture (SOA), Web services are based on a number of widely acknowledged design principles, qualities and structural features that are different from traditional systems [2,20,29,30]. While there is no generalized recipe for what is considered to be a good service design, there exists guidelines about how to develop service-oriented designs while following a set of quality principles like service reusability, flexibility, and loose coupling principles [10,20,29]. However, like any software system, Web service must evolve to add new user requirements, fix defects or adapt to new environment changes. Such frequent changes, as well as other business factors, developers expertise, and deadline pressure may, in turn, lead to the violation of design quality principles. The existence of bad programming practices, inducing poor design, also called *"anti-patterns"* or *"design defects"*, are an indication of such violations [20,21,25]. Such antipatterns include the *God Object Web service* which typically refers to a Web service with large interface implementing a multitude of methods related to different technical and business abstractions. The *God Object Web Service* is not easy to discover and reuse and often unavailable to end users because it is overloaded [12]. Moreover, when many clients are utilizing one interface, and several developers work on one underlying implementation, there are bound to be issues of breakage in clients and developer contention for changes to server-side implementation artifacts [12]. To this end, such anti-patterns should be detected, prevented and fixed in real world SBS to adequately fit in the required system's design with high QoS [12,29].

While recent works attempted to detect and fix Web service antipatterns [18–21,25,33], the detection of such antipatterns is still a challenging and difficult task. Indeed, there is no consensual way to translate formal definition and symptoms into actionable detection rules. Some efforts attempted to manually define detection rules [18,19,25]. However, such manual rules are applied, in general, to a limited scope and require a non-trivial manual effort and human expertise to calibrate a set of detection rules to match the symptoms of each antipattern instance with the actual characteristics of a given Web service. Other approaches attempted to use machine learning to better automate the detection of antipatterns [20,21,32]. However, these detection approaches formulated to the detection as a single-label learning problem, *i.e.*, dealing with antipatterns independently and thus ignoring the innate relationships between the different antipattern symptoms and characteristics. As a result, existing machine learning-based approaches lead to several false positives and true negatives reducing the detection accuracy. Indeed, recent studies showed that different types of Web service antipatterns may exhibit similar symptoms and can thus co-exist in the same Web service [12,20,25]. That is, similar symptoms can be used to characterize multiple antipattern types making their identification even harder and error-prone [12,20]. For example, the *God Object Web service* (GOWS) antipattern is typically associated with the *Chatty Web service* (CWS) antipattern which manifests in the form of a Web service with a high number of operations that are required to complete abstraction. Consequently, the GOWS and the CWS typically co-occur in Web services. Inversely, the GOWS antipattern has different symptoms than the fine-grained Web service (FGWS) antipattern which typically refers to a small Web service with few operations implementing only apart of an abstraction. Hence, knowing that a Web service is detected as a FGWS antipattern, it cannot be a GOWS or a CWS as they have different innate characteristics/symptoms.

In this paper, our aim is to provide an automated and accurate technique to detect Web service anti-patterns. We formulate the Web services antipatterns detection problem as a multi-label learning (MLL) problem to deal with the interleaving symptoms of existing Web service antipatterns by generating multiple detection rules that can detect various antipattern types. We use the ensemble classifier chain (ECC) technique [28] that converts the detection task of multiple antipattern types into several binary classification problems for each individual antipattern type. ECC involves the training of n single-label binary classifiers, where each one is solely responsible for detecting a specific label, *i.e.*, antipattern type. These n classifiers are linked in a chain, such that each binary classifier is able to consider the labels identified by the previous ones as additional information at the classification time. For the binary classification, we exploit the effectiveness of genetic programming (GP) [13,14,20] to find the optimal detection rules for each antipattern. The goal of GP is to learn detection rules from a set of real-world instances of Web service antipatterns. In fact, we use GP to translate regularities and symptoms that can be found in real-world Web service antipattern examples into actionable detection rules. A detection rule is a combination of Web service interface quality metrics with their appropriate threshold values to detect various types of antipatterns.

We implemented and evaluated our approach on a benchmark of 815 Web services from different application domains and sizes and eight common Web service antipattern types. To evaluate the performance of our GP-ECC approach, and the statistical analysis of our results show that the generated detection rules can identify the eight considered antipattern types with an average precision of 89%, and recall of 93% and outperforms state-of-the-art techniques [20,25]. Moreover, we conducted a deep analysis to investigate the symptoms, *i.e.*, features, that are the best indicators of antipatterns. We found that the most influential factors that best characterize Web service anti-patterns include the number of declared operations, the number of port types, and the number of simple and complex types in service interfaces.

This paper is structured as follows: the paper's background is detailed in Sect. 2. Section 3 summarizes the related studies. In Sect. 4, we describe our GP-ECC approach for Web service antipatterns detection. Section 5 presents our experimental evaluation, and discusses the obtained results. Section 6 discusses potential threats to the validity our our approach. Finally, Sect. 7 concludes and outlines our future work.

2   Background

This section describes the basic concepts used in this paper.

2.1   Web Service Anti-Patterns

Anti-patterns are symptoms of bad programming practices and poor design choices when structuring the web interfaces.They typically engender web interfaces to become harder to maintain and understand [17].

Various types of antipatterns, characterized by how they hinder the quality of service design, have been recently introduced with the purpose of identifying them, in order to suggest their removal through necessary refactorings [12,16,25]. Typical web service antipatterns are described in Table 1:

Table 1. The list of considered Web service antipatterns.

Antipatterns definitions
Chatty Web service (CWS): is a service where a high number of operations, typically attribute-level setters or getters, are required to complete one abstraction. This antipattern may have many fine-grained operations, which degrades the overall performance with higher response time [12]
Fine grained Web service (FGWS): is a too fine-grained service whose overhead (communications, maintenance, and so on) outweighs its utility. This defect refers to a small Web service with few operations implementing only a part of an abstraction. It often requires several coupled Web services to complete an abstraction, resulting in higher development complexity, reduced usability [12]
God object Web service (GOWS): implements a multitude of methods related to different business and technical abstractions in a single service. It is not easily reusable because of the low cohesion of its methods and is often unavailable to end users because it is overloaded [12]
Ambiguous Web service (AWS): is an antipattern where developers use ambiguous or meaningless names for denoting the main elements of interface elements (e.g., port types, operations, messages). Ambiguous names are not semantically and syntactically sound and affect the service discoverability and reusability [19]
Data Web service (DWS): contains typically accessor operations, i.e., getters and setters. In a distributed environment, some Web services may only perform some simple information retrieval or data access operations. A DWS usually deals with very small messages of primitive types and may have high data cohesion [25]
CRUDy Interface (CI): is a service with RPC-like behavior by declaring create, read, update, and delete (CRUD) operations, e.g., `createX()`, `readY()`, etc. Interfaces designed in that way might be chatty because multiple operations need to be invoked to achieve one goal. In general, CRUD operations should not be exposed via interfaces [12]
Redundant PortTypes (RPT): is a service where multiple portTypes are duplicated with the similar set of operations. Very often, such portTypes deal with the same messages. RPT antipattern may negatively impact the ranking of the Web Services [12]
Maybe It is Not RPC (MNR): is an antipattern where the Web service mainly provides CRUD- type operations for significant business entities. These operations will likely need to specify a significant number of parameters and/or complexity in those parameters. This antipattern causes poor system performance because the clients often wait for the synchronous responses [12]

We focus our study on these eight antipattern types as they are the most common ones in SBSs based on recent studies [15,16,21,22,25,30,33].

2.2 Multi-label Learning

Multi-label learning (MLL) is the machine learning task of automatically assigning an object into multiple categories based on its characteristics [5,28,31]. Single-label learning is limited by one instance with only one label. MLL is a nontrivial generalization by removing the restriction and it has been a hot topic in machine learning [5]. MLL has been explored in many areas in machine learning and data mining fields through classification techniques. There exists different MLL techniques [3,5,28,31,35] including (1) problem transformation methods and algorithms, *e.g.*, the classifier chain (CC) algorithm, the binary relevance (BR) algorithm, label powerset (LP) algorithm, and (2) algorithm adaptation methods such as the K-Nearest Neighbors (ML.KNN), as well as (3) ensemble methods such as the ensemble classifier chain (ECC), and random k-labelset (RAKEL).

The Classifier Chain (CC) Model. The CC model combines the computational efficiency of the BR method while still being able to take the label dependencies into account for classification. With BR, the classifier chains method involves the training of q single-label binary classifiers and each one will be solely responsible for classifying a specific label $l_1, l_2, ..., l_q$. The difference is that, in CC, these q classifiers are linked in a chain $\{h_1 \rightarrow h_2 \rightarrow ... \rightarrow h_q\}$ through the feature space. That is, during the learning time, each binary classifier h_j incorporates the labels predicted by the previous $h_1, ..., h_{j-1}$ classifiers as additional information. This is accomplished using a simple trick: in the training phase, the feature vector x for each classifier h_j is extended with the binary values of the labels $l_1, ..., l_{j-1}$.

The Ensemble Classifier Chain (ECC) Model. One of the limitation of the CC model is that the order of the labels is random. This can lead may lead to a single standalone CC model be poorly ordered. Moreover, there is the possible effect of error propagation along the chain at classification time, when one (or more) of the first classifiers predict poorly [28]. Using an ensemble of chains, each with a random label order, greatly reduces the risk of these events having an overall negative effect on classification accuracy. A majority voting method is used to select the best model. Moreover, a common advantage of ensembles is their performance in increasing overall predictive performance [3,28].

2.3 Genetic Programming

Genetic Programming (GP) [13], a sub-family of Genetic Algorithms (GA), is a computational paradigm that were inspired by the mechanics of natural evolution, including survival of the fittest, reproduction, and mutation. GP begins with a set of random population of candidate solutions, also called individuals or chromosomes. Each individual of the population, is represented in the form of a computer program or tree and evaluated by a fitness function to quantitatively measure its ability to solve the target problem.

In this paper, we apply GP to the problem of Web service antipatterns detection. Hence, we show how GP can effectively explore a large space of solutions, and provide intelligible detection rules with ECC. Also, we bridge the gap between MLL and GP based on the ECC method to solve the problem of antipatterns detection, where each Web service may contain different interleaving antipatterns, *e.g.*, GOWS, CWS and CI. For the binary labels, our ECC model adopts GP to learn detection rules for each antipattern type.

3 Related Work

Detecting and specifying antipatterns in SOA and Web services is a relatively new field. The first book in the literature was written by Dudney et al. [12] and provides informal definitions of a set of Web service antipatterns. More recently, Rotem-Gal-Oz described the symptoms of a range of SOA antipatterns [30]. Furthermore, Král et al. [16] listed seven "popular" SOA antipatterns that violate accepted SOA principles. In addition, a number of research works have addressed the detection of such antipatterns. Recently, Palma et al. [25] have proposed a rule-based approach called SODA-W that relies on declarative rule specification using a domain-specific language (DSL) to specify/identify the key symptoms that characterize an antipattern using a set of WSDL metrics. In another study, Rodriguez et al. [29] and Mateos et al. [19] provided a set of guidelines for service providers to avoid bad practices while writing WSDLs. Based on some heuristics, the authors detected eight bad practices in the writing of WSDL for Web services. Mateos et al. [18] have proposed an interesting approach towards generating WSDL documents with less antipatterns using text mining techniques. Ouni et al. [20,22] proposed a search-based approach based on evolutionary computation to find regularities, from examples of Web service antipatterns, to be translated into detection rules. However, detections rules based approaches tend to have a higher number of false positives. Ouni et al. [21] introduced a machine learning based approach to build detection models for different Web service antipattern types. However, the major limitation of the current approaches is that deal with the Web service antipatterns problem as a single label learning problem ignoring the valuable information related to the shared symptoms between different antipattern types. As a consequence, they suffer from reduced accuracy related to several false positives and true negatives.

To fix such antipatterns, Daagi et al. [9] proposed an automated approach based on formal concept analysis to fix the GOWS antipattern. Ouni et al. [23,24] introduced a hybrid approach based on graph partitioning and search based optimization to improve the design quality of web service interfaces to reduce coupling and increase cohesion. Later, Wang et al. [33] have formulated an interactive approach to find the optimal design of Web service and reduce the number of antipatterns.

4   Approach

In this section, we provide the problem formulation for Web service antipatterns detection as a MLL problem. Then, we describe the details of our approach.

4.1   Problem Formulation

We define the Web services antipatterns detection problem as a multi-label learning problem. Each antipattern type is denoted by a label l_i. A MLL problem can be formulated as follows. Let $X = R^d$ denote the input feature space. $L = \{l_1, l_2, ...l_q\}$ denote the finite set of q possible labels, *i.e.*, antipattern types. Given a multi-label training set $D = \{(x_1, y_1), (x_2, y_2), ...(x_N, y_N)\}(x_i \in X, y_i \subseteq L)$, the goal of the multi-label learning system is to learn a function $h : X \to 2^L$ from D which predicts a set of labels for each unseen instance based on a set of known data.

4.2   Approach Overview

The main goal of our approach is to generate a set of detection rules for each antipattern type while taking into consideration the dependencies between the different antipatterns and their interleaving symptoms. Figure 1 presents an overview of our approach to generate service antipatterns detection rules using the GP-ECC model. Our approach consists of two phases: *training* phase and *detection* phase. In the training phase, our goal is to build an ensemble classifier chain (ECC) model learned from real-world antipattern instances identified from existing Web services based on several GP models for each individual antipattern. In the detection phase, we apply this model to detect the proper set of labels (*i.e.*, antipattern type) for a new unlabeled data (*i.e.*, a new Web service).

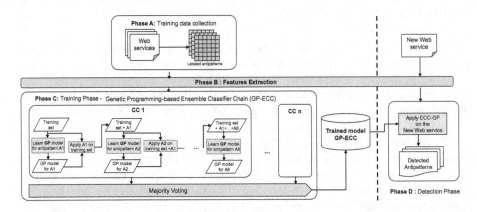

Fig. 1. The Web service antipatterns detection framework using GP-ECC.

Our approach takes as inputs a set of Web services with known labels, *i.e.*, antipatterns (phase A). Then, extracts a set of features characterizing the considered antipattern types from which a GP algorithm will learn (phase B). Next, an ECC algorithm will be built (phase C). The ECC algorithm consists of a set of classifier chain models (CC), each with a random label order. Each CC model, learns eight individual GP models for each of the eight considered antipattern types. The ith binary GP detector will learn from the training data while considering the existing i already detected antipatterns by the $i-1$ detected antipatterns to generate the optimal detection rule that can detect the current i^{th} antipattern. In total, the ECC trains n multi-label CC classifiers $CC_1, ..., CC_n$; each classifier is given a random chain ordering; each CC builds 8 binary GP models for each antipattern type. Each binary model uses the previously predicted binary labels into its feature space. Then, our framework searches for the near optimal GP-ECC model from these n multi-label chain classifiers using an ensemble majority voting schema based on each label confidence [28].

In the detection phase, the returned GP-ECC model is a machine learning classifier that assigns multiple labels, *i.e.*, antipattern types, to a new Web service based on its current features, *i.e.*, its symptoms (phase D). In the next subsections, we provide the details of each phase.

4.3 Phase A : Training Data Collection

Our proposed technique leverages knowledge from a set of examples containing real world instances of web service antipatterns. The base of examples contains different web service antipatterns from different application domains (*e.g.*, social media, weather, online shopping, etc.), which were gathered from various Web service online repositories and search engines, like ProgrammableWeb, and ServiceXplorer, etc. To ensure the correctness of our dataset, the studied antipattern instances were manually inspected and verified according to existing guidelines from the literature [12,16,20,25]. Our dataset is publicly available [1].

4.4 Phase B : Features Extraction Module

The proposed techniques leverages a set of popular and widely used metrics related to web services [20–22,25,27,32,34]. As shown in Table 2, our technique develops its detection rules using suite of over 42 quality metrics including (*i*) code level metrics, (*ii*) WSDL interface metrics, and (*iii*) measurements of performance. Code metrics are calculating using service Java artefacts, being mined using the Java[TM] API for XML Web Services (JAX-WS)[1] [8] as well as the *ckjm* tool[2] (Chidamber & Kemerer Java Metrics) [6]. WSDL metrics capture any design properties of Web services, in the structure of the WSDL interface level. Furthermore, a set of dynamic metrics is also captured, using web service invocations, *e.g.*, availability, and response time.

---

[1] http://docs.oracle.com/javase/6/docs/technotes/tools/share/wsimport.html.

[2] http://gromit.iiar.pwr.wroc.pl/p_inf/ckjm/.

4.5 Phase C : Multi-label Learning Using GP-ECC

As outlined in the previous Sect. 4.2, our solution uses the ECC classifier [28] to model the multi-label learning task into multiple single-label learning tasks. Our multi-label ECC is a detection model for the identification of web service antipatterns. Each classifier chain (CC) represents a GP model (binary decision tree) for each smell type while considering the previously detected smells (if any), *i.e.*, each binary GP model uses the previously predicted binary labels into its feature space. The motivation behind the choice of GP-based models is driven by its efficiency in the resolution of similar software engineering problems such as, design defects, and code smells identification [3,14,20,22].

In our approach, we adopted the Non-dominated Sorting Genetic Algorithm (NSGA-II) [11] as a search algorithm to generate antipatterns detection rules. NSGA-II is a powerful and widely-used evolutionary algorithm which extends the generic model of GP learning to the space of programs. Unlike other evolutionary search algorithms, in our NSGA-II adaptation, solutions are themselves programs following a tree-like representation instead of fixed length linear string formed from a limited alphabet of symbols [13]. More details about NSGA-II can be found in Deb et al. [11].

We describe in the following subsections the three main adaptation steps: (*i*) solution representation, (*ii*) the generation of the initial generation (*iii*) fitness function, and (*iv*) change operators.

(*i*) **Solution Representation.** A solution consists of a rule that can detect a specific type of anti-pattern in the form of:

IF (Combination of metrics and their thresholds) **THEN** antipattern type.

In MOGP, the solution representation is a tree-based structure of functions and terminals. Terminals represent various structural, dynamic, and service oriented metrics, extracted from Table 2. Functions are logic operators such as OR (union), AND (intersection), or XOR (eXclusive OR). Thus, each solution is encoded as a binary tree with leafnodes (terminals) correspond to one of metrics described in Table 2 and their associated threshold values randomly generated. Internal-nodes (functions) connect sub-tress and leaves using the operators set $C = \{AND, OR, XOR\}$. Figure 2 is a simplified illustration of a given solution.

(*ii*) **Generation of the Initial Population.** The initial population of solutions is generated randomly by assigning a variety of metrics and their thresholds to the set of different nodes of the tree. The size of a solution, *i.e.*, the tree's length, is randomly chosen between lower and upper bound values. These two bounds have determined and called the problem of bloat control in GP, where the goal is to identify the tree size limits. Thus, we applied several trial and error experiments using the HyperVolume (HP) performance indicator [13] to determine the upper bound after which, the sign remains invariant.

(*iii*) **Fitness Function.** The fitness function evaluates how good is a candidate solution in detecting web service antipatterns. Thus, to evaluate the fitness

Table 2. List of Web service quality metrics used.

Metric	Description	Metric level
Service interface metrics		
NPT	Number of port types	Port type
NOD	Number of operations declared	Port type
NCO	Number of CRUD operations	Port type
NOPT	Average number of operations in port types	Port type
NPO	Average number of parameters in operations	Operation
NCT	Number of complex types	Type
NAOD	Number of accessor operations declared	Port type
NCTP	Number of complex type parameters	Type
COUP	Coupling	Port type
COH	Cohesion	Port type
NOM	Number of messages	Message
NST	Number of primitive types	Type
ALOS	Average length of operations signature	Operation
ALPS	Average length of port types signature	Port type
ALMS	Average length of message signature	Message
RPT	Ratio of primitive types over all defined types	Type
RAOD	Ratio of accessor operations declared	Port type
ANIPO	Average number of input parameters in operations	Operation
ANOPO	Average number of output parameters in operations	Operation
NPM	Average number of parts per message	Message
AMTO	Average number of meaningful terms in operation names	Operation
AMTM	Average number of meaningful terms in message names	Message
AMTP	Average number of meaningful terms in port type names	Type
Service code metrics		
WMC	Weighted methods per class	Class
DIT	Depth of Inheritance Tree	Class
NOC	Number of Children	Class
CBO	Coupling between object classes	Class
RFC	Response for a Class	Class
LCOM	Lack of cohesion in methods	Class
Ca	Afferent couplings	Class
Ce	Efferent couplings	Class
NPM	Number of Public Methods	Class
LCOM3	Lack of cohesion in methods	Class
LOC	Lines of Code	Class
DAM	Data Access Metric	Class
MOA	Measure of Aggregation	Class
MFA	Measure of Functional Abstraction	Class
CAM	Cohesion Among Methods of Class	Class
AMC	Average Method Complexity	Method
CC	The McCabe's cyclomatic complexity	Method
Service Performance Metrics		
RT	Response Time	Method
AVL	Availability	Service

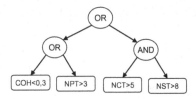

Fig. 2. A simplified example of a solution for GOWS antipattern.

of each solution, we use two objective functions, based on two well-known metrics [14,20], to be optimized, *i.e.*, precision and recall. The precision objective
function aims at maximizing the detection of correct antipatterns over the list of
detected ones. The recall objective function aims at maximizing the coverage of
expected antipatterns from the base of examples over the actual list of detected
instances. Precision and recall of a solution S are defined as follows.

$$Precision(S) = \frac{|\{\text{Detected antipatterns}\} \cap \{\text{Expected antipatterns}\}|}{|\{\text{Detected antipatterns}\}|} \quad (1)$$

$$Recall(S) = \frac{|\{\text{Detected antipatterns}\} \cap \{\text{Expected antipatterns}\}|}{|\{\text{Expected antipatterns}\}|} \quad (2)$$

(*iv*) **Change Operators.** Crossover and mutation are used as change operators to evolve candidate solutions towards optimality.

Crossover. We adopt the "standard" random, single-point crossover. It selects
two parent solutions at random, then picks a sub-tree on each one. Then, the
crossover operator swaps the nodes and their relative subtrees from one parent
to the other. Each child thus combines information from both parents.

Mutation. The mutation operator aims at slightly changing a solution, without losing its main properties. It can be applied for both function or terminal
nodes. For a given solution to mutated, one or multiple nodes are randomly
selected, then changed according to their type. For functions, a logic operator
can be replaced with any randomly selected logic operator, while for terminals,
metrics may be swapped with another metric, or a metric threshold can be
changed.

ECC Majority Voting. As shown in Fig. 1, for each CC, MOGP will generate
an optimal rule for each type of web service antipattern, *i.e.*, binary detection.
Then, ECC allows to find the best CC that provides the best MLL from all
the trained binary models. Each CC_i model is likely to be unique and able to
achieve different multi-label classifications. These classifications are summed by
label so that each label receives a number of votes. A threshold is used to select
the most popular labels which form the final predicted multi-label set. This is a
generic voting scheme used in MLL transformation methods [28].

4.6   Phase D: Detection Phase

After constructing the GP-ECC model in the training phase, it will be then used
to detect a set of labels for a new Web service. It takes as input the set of features

extracted from a given Web service using the feature extraction module. As output, it returns the detection results for each individual label, *i.e.*, antipattern.

5   Empirical Study

In this section, we describe our empirical study to evaluate our GP-ECC approach. We report the research questions, the experimental setup, and results.

5.1   Research Questions

We designed our empirical study to answer the three following research questions.

- **RQ1: (Performance)** How accurately can our GP-ECC approach detect Web service anipatterns?
- **RQ2: (Sensitivity)** What types of anipatterns does our GP-ECC approach detect correctly?
- **RQ3: (Features influence)** What are the most influential features that can indicate the presence of anipatterns?

5.2   Experimental Setup

We evaluate our approach on a benchmark of 815 Web services [1]. Table 3 summarizes the experimental dataset. Furthermore, as a sanity check, all antipatterns were manually inspected and validated based on literature guidelines [12,30] as discussed in Sect. 4.3. Web services were collected from different Web service search engines including eil.cs.txstate.edu/ServiceXplorer, programmableweb.com, biocatalogue.org, webservices.seekda.com, taverna.org.uk and myexperiment.org. Furthermore, for better generalisability, our empirical study, our collected Web services are drawn from 9 different application domains, *e.g.*, financial, science, search, shipping, etc.

We considered eight common types of Web service antipatterns, *i.e.*, *god object Web service* (GOWS), *fine-grained Web service* (FGWS), *chatty Web service* (CWS), *data Web service* (DWS), *ambiguous Web service* (AWS), *redundant port types* (RPT), *CRUDy interface* (CI), and *Maybe It is Not RPC* (MNR), (cf. Sect. 2). In our experiments, we conducted a 10-fold cross-validation procedure to split our data into training data and evaluation data.

To answer **RQ1**, we conduct experiments to justify our GP-ECC approach.

Baseline Learning Methods. We first compare the performance of our meta-algorithm ECC. We used GP, *decision tree* (J48) and *random forest* (RF) as corresponding basic classification algorithms. We also compared with the widely used MLL algorithm adaptation method, *K-Nearest Neighbors* (ML.KNN). Thus, in total, we have 4 MLL algorithms to be compared. One fold is used for the test and 9 folds for the training.

Table 3. The list of Web services used in our evaluation.

Category	# of services	# of antipatterns
Financial	185	115
Science	52	18
Search	75	33
Shipping	58	23
Travel	105	49
Weather	65	21
Media	82	19
Education	55	28
Messaging	63	43
Location	75	39
All	**815**	**388**

State-of-the-Art Detection Methods. Moreover, we compare the performance of our approach with two state-of-the-art approaches, SODA-W [25] and P-EA [20] for Web service antipattern detection. The SODA-W approach of Palma et al. [25] manually translates antipattern symptoms into detection rules and algorithms based on a literature review of Web service design. P-EA [20] adopts parallel GP technique to detect Web service antipatterns based on a set of Web service interface metrics. Both approaches detect antipattern types in an independent manner.

To compare the performance of each method, we use common performance metrics, *i.e.*, precision, recall, and F-measure [14,20,28]. Let l a label in the label set L. For each instance i in the antipatterns learning dataset, there are four outcomes, True Positive (TP_l) when i is detected as label l and it correctly belongs to l; False Positive (FP_l) when i is detected as label l and it actually does not belong to l; False Negative (FN_l) when i is not detected as label l when it actually belongs to l; or True Negative (TN_l) when i is not detected as label l and it actually does not belong to l. Based on these possible outcomes, precision (P_l), recall (R_l) and F-measure (F_l) for label l are defined as follows:

$$P_l = \frac{TP_l}{TP_l + FP_l} \quad ; \quad R_l = \frac{TP_l}{TP_l + FN_l} \quad ; \quad F_l = \frac{2 \times P_l \times R_l}{P_l + R_l}$$

Then, the average precision, recall, and F-measure of the $\mid L \mid$ labels are calculated as follows:

$$Precision = \frac{1}{\mid L \mid} \sum_{l \in L} P_l \quad ; \quad Recall = \frac{1}{\mid L \mid} \sum_{l \in L} R_l \quad ; \quad F1 = \frac{1}{\mid L \mid} \sum_{l \in L} F_l$$

Statistical Test Methods. To compare the performance of each method, we perform Wilcoxon pairwise comparisons [7] at 99% significance level (*i.e.*, $\alpha =$

0.01) to compare GP-ECC with each of the 9 other methods. We also used the non-parametric effect Cliff's delta (d) [7] to compute the effect size. The effect size d is interpreted as Negligible if $\mid d \mid < 0.147$, Small if $0.147 \leq\mid d \mid < 0.33$, Medium if $0.33 \leq\mid d \mid < 0.474$, or High if $\mid d \mid \geq 0.474$.

To answer **RQ2**, we investigated the antipattern types that were detected to find out whether there is a bias towards the detection of specific types.

To answer **RQ3**, we aim at identifying the features that are the most important indicators of whether a Web service has a given antipattern or not. For each antipattern type, we count the percentage of rules in which the feature appears across all obtained optimal rules by GP. The more a feature appears in the set of optimal trees, the more the feature is relevant to characterize that antipattern.

5.3   Results

Results for RQ1 (Performance). Table 4 reports the average precision, recall and F-measure scores for the compared methods. We observe that we see that ECC performs well with GP as a base method as compared to J48 and RF. We used GP-ECC as the base for determining statistical significance. In particular, the GP-ECC method achieves the highest F-measure with 0.91 compared to the J48 and RF methods achieving an F-measure of 0l9 and 0.89, respectively, with medium and large effect sizes. The same performance was achieved by GP-ECC in terms of precision and recall, with 0.89 and 0.93, respectively. The statistical analysis of the obtained results confirms thus the suitability of the GP formulation compared to decision tree and random forest algorithms. We can also see overall superiority for of the ECC and in particular the GP-ECC compared to the transformation method ML.KNN in terms of precision, recall and F-measure with large effect size. One of the reasons that ML.KNN does not perform well is that it ignores the label correlation, while ECC consider the label correlation by using an ensemble of classifiers. Moreover, among the 3 base learning algorithms, GP performs the best, followed by J48 and RF.

Moreover, we observe from Table 4 that GP-ECC achieved a higher superiority than both state-of-the-art approaches, P-EA and SODA-W. While P-EA achieves promising results with an average F-measure of 83%, it is still less than GP-ECC. Moreover, SODA-W achives an F-measure of 72% which lower than other approaches. We conjecture that a key problem with P-EA and SODA-W is that they detect separately possible antipatterns without looking at the relationship between them. Through a closer manual inspection of the false positive and true negative instances by P-EA and SODA-W, we found a number of Web services that are detected at the same time as god object Web services (GOWS) and fine-grained Web services (FGWS) which would reduce the overall accuracy as GOWS and FGWS cannot co-occur in the same Web service. Other missing chatty Web service (CWS) instances were identified in Web services that are detected as GOWS. Indeed, GP-ECC makes the hidden relationship between antipatterns more explicit which has shown higher accuracy.

Table 4. The achieved results by ECC with the base algorithms GP, J48, and RF; ML.KNN; and existing approaches SODA-W and P-EA.

Approach	Precision		Recall		F1	
	Score	p-value (d)*	Score	p-value (d)*	Score	p-value (d)*
GP-ECC	**0.89**	-	**0.93**	-	**0.91**	-
J48-ECC	0.87	<0.01 (M)	0.9	<0.01 (M)	0.88	<0.01 (M)
RF-ECC	0.86	<0.01 (M)	0.89	<0.01 (M)	0.87	<0.01 (M)
ML.KNN	0.83	<0.01 (L)	0.84	<0.01 (L)	0.83	<0.01 (L)
P-EA	0.82	<0.01 (L)	0.85	<0.01 (L)	0.83	<0.01 (L)
SODA-W	0.7	<0.01 (L)	0.74	<0.01 (L)	0.72	<0.01 (L)

*p-value(d) reports the statistical difference (p-value) and effect-size (d) between GP-ECC and the algorithm/approach in the current row.
The effect-size (d) is N : Negligible − S : Small − M : Medium − L : Large

Results for RQ2 (Sensitivity). Figure 3 reports the sensitivity analysis of each specific antipattern type. We observe that GP-ECC does not have a bias towards the detection of any specific antipattern type. As shown in the figure, GP-ECC achieved good performance and low variability in terms of the median F-measure, ranging from 87% to 93%, across the 8 considered antipattern types. The highest F-measure was obtained for the god object (GOWS) and fine-grained (FGWS) antipatterns (93%) which heavily relies on the notion of size. This higher performance is reasonable since the existing guidelines [12,30] rely heavily on the notion of size in terms of declared operations, port types, and simple/complex data types used. But for antipatterns such as the ambiguous Web service (AWS), the notion of size is less important, it rather relies an the meaningfulness and length of operations and messages identifiers. This aspect makes this type of antipatterns hard to detect using such information as it often depends on human interpretations.

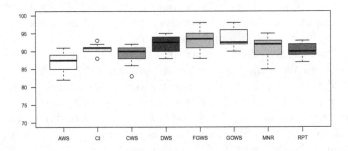

Fig. 3. F-measure achieved by GP-ECC for each antipattern across all categories.

Results for RQ3 (Features Influence). To better understand what features, *i.e.*, metrics, are most used by our GP-ECC model to generate detection rules among all the generated rules, we count the percentage of rules in which the feature appears. Table 5 shows the statistics for each smell type with the top-10 features (cf. Table 2), from which the three most influencing features values are in bold. We observe that the number of operations declared (NOD), the number of messages (NOM), the number of simple and complex types (NST and NCT), and the cohesion (COH) are the most influencing parameters. Other features such as the number of input parameters in operations (NIPO) and the coupling (COUP) are also influencing the existence of antipatterns. We also found that some features such the average length of operation signatures (ALOS) are specific to the ambiguous Web service (AWS) antipattern and do not participate to characterize any of the other considered antipattern types.

Table 5. The most influential features for each antipattern.

	GOWS	FGWS	CWS	DWS	AWS	RPT	CI	MNR
NOD	**98**	**100**	91	86	52	**96**	**93**	83
NOM	**92**	**90**	100	92	55	**91**	89	**93**
COH	**89**	84	89	87	23	**92**	**92**	85
WMC	82	82	81	65	45	81	72	83
NIPO	79	75	89	84	**92**	61	**93**	**98**
NCT	81	85	91	93	41	54	86	91
NST	**89**	**86**	**96**	**96**	**82**	32	80	93
ALOS	34	32	41	18	**100**	9	39	23
COUP	76	69	**93**	82	71	52	89	**100**
LCOM	88	77	88	85	46	81	86	79

We thus observe that different interface service level measures play a crucial role in the emergence of antipatterns, while those related to the source code are less influencial. These findings suggest that more attention has to be paid to the design of their service interface to avoid the presence of antipatterns and their impact on the software quality. This finding aligns also with previous research advocating the importance of service interface design [4, 12, 18, 24, 26, 33]

6 Threats to Validity

Threats to construct validity could be related to the performance measures. We basically used standard performance metrics such as precision, recall and F-measure that are widely accepted in MLL and software engineering [20, 25]. Another potential threat could be related to the selection of learning techniques. Although we use the GP, J48 and RF which are known to have high performance, we plan to compare with other ML techniques in our future work.

Threats to internal validity relate to errors in our experiments. Our approach relies on the used metrics to characterize antipatterns. We mitigated this issues by using popular and well-accepted metrics and tools to neasure our metrics.

Threats to external validity relate to the generalizability of our results. Our approach relies on learning from existing services, and so, their diversity is critical for our learning process. We mitigated this threat by choosing independent services, issued from different providers, and they were also developed in multiple application domains. Also, our training set was manually validated, however, such human activity is prone to error sand personal bias. The reduction of such bias can be achieved by following existing literature gidelines [12,16] randomly choosing a statistically significant sample that is reclassified by the three authors. Then, the kappa agreement is calculated and its corresponding score is 0.83, which is considered a high score for inter-rater agreement [7].

7   Conclusion and Future Work

Web service antipatterns are symptoms of potential problems threatening the longevity of services. Although such antipatterns can facilitate the coding the quick delivery of services, their long-term impact hinders the maintainability and evolvability of services. This paper developed a novel technique, leveraging an existing set of manually verified antipatterns, to develop a metric-based detection rules using ensemble classifier chain. We transform multi-label problems into several single-label problems that are solved using the genetic programming. Our experiments show the effectiveness of our detection strategy by achieving an F-Measure of 93%, when analyzing a large set of 815 web services.

As part of our future investigations, we plan on extending the set of metrics we used as well as other RESTFul Web services, in order to explore potential features, which may further improve the accuracy of our detection strategy.

References

1. Replication package (2020). https://github.com/WS-antipatterns/dataset
2. Almarimi, N., Ouni, A., Bouktif, S., Mkaouer, M.W., Kula, R.G., Saied, M.A.: Web service api recommendation for automated mashup creation using multi-objective evolutionary search. Appl. Soft Comput. **85**, 105830 (2019)
3. Almarimi, N., Ouni, A., Chouchen, M., Saidani, Islem, M.M.W.: On the detection of community smells using genetic programming-based ensemble classifier chain. In: International Conference on Global Software Engineering, pp. 1–12 (2020)
4. Boukharata, S., Ouni, A., Kessentini, M., Bouktif, S., Wang, H.: Improving web service interfaces modularity using multi-objective optimization. Automated Softw. Eng. **26**(2), 275–312 (2019). https://doi.org/10.1007/s10515-019-00256-4
5. de Carvalho, A.C.P.L.F., Freitas, A.A.: A Tutorial on Multi-label Classification Techniques, pp. 177–195 (2009)
6. Chidamber, S.R., Kemerer, C.F.: A metrics suite for object oriented design. IEEE Trans. Softw. Eng. **20**(6), 476–493 (1994)

7. Cohen, J.: Statistical power analysis for the behavioral sciences. Academic Press (1988)
8. Coscia, J.L.O., Crasso, M., Mateos, C., Zunino, A.: Estimating web service interface quality through conventional object-oriented metrics. CLEI E. **16**(1) 2056–2101 (2013)
9. Daagi, M., Ouni, A., Kessentini, M., Gammoudi, M.M., Bouktif, S.: Web service interface decomposition using formal concept analysis. In: IEEE International Conference on Web Services (ICWS), pp. 172–179 (2017)
10. Daigneau, R.: Service Design Patterns: fundamental design solutions for SOAP/WSDL and restful Web Services. Addison-Wesley (2011)
11. Deb, K., Pratap, A., Agarwal, S., Meyarivan, T.: A fast and elitist multiobjective genetic algorithm: NSGA-II. IEEE Trans. Evol. Comput. **6**(2), 182–197 (2002)
12. Dudney, B., Krozak, J., Wittkopf, K., Asbury, S., Osborne, D.: J2EE Antipatterns. Wiley, Hoboken (2003)
13. John, R., Koza, M.: Genetic programming: On programming computers by means of natural selection and genetics. In: Association for Computing Machinery, MIT Press, Cambridge (1992)
14. Kessentini, M., Ouni, A.: Detecting android smells using multi-objective genetic programming. In: IEEE/ACM 4th International Conference on Mobile Software Engineering and Systems (MOBILESoft), pp. 122–132 (2017)
15. Král, J., Žemlička, M.: Crucial service-oriented antipatterns. Int. J. Adv. Softw. **2**(1), 160–171 (2009)
16. Král, J., Zemlicka, M.: Popular SOA Antipatterns. In: Computation World: Future Computing, Service Computation, Cognitive, Adaptive, Content, Patterns, pp. 271–276 (2009)
17. Marinescu, R.: Detection strategies: metrics-based rules for detecting design flaws. In: 2013 IEEE International Conference on Software Maintenance, pp. 350–359 (2004)
18. Mateos, C., Rodriguez, J.M., Zunino, A.: A tool to improve code-first web services discoverability through text mining techniques. Softw. Pract. Experience **45**(7), 925–948 (2015)
19. Mateos, C., Zunino, A., Coscia, J.L.O.: Avoiding WSDL bad practices in code-first web services. SADIO Electron. J. Inform. Oper. Res. **11**(1), 31–48 (2012)
20. Ouni, A., Kessentini, M., Inoue, K., Cinneide, M.O.: Search-based web service antipatterns detection. IEEE Trans. Serv. Comput. **10**(4), 603–617 (2017)
21. Ouni, A., Daagi, M., Kessentini, M., Bouktif, S., Gammoudi, M.M.: A machine learning-based approach to detect web service design defects. In: IEEE International Conference on Web Services (ICWS). pp. 532–539 (2017)
22. Ouni, A., Gaikovina Kula, R., Kessentini, M., Inoue, K.: Web service antipatterns detection using genetic programming. In: Annual Conference on Genetic and Evolutionary Computation (GECCO), pp. 1351–1358 (2015)
23. Ouni, A., Salem, Z., Inoue, K., Soui, M.: SIM: an automated approach to improve web service interface modularization. In: IEEE International Conference on Web Services (ICWS), pp. 91–98 (2016)
24. Ouni, A., Wang, H., Kessentini, M., Bouktif, S., Inoue, K.: A hybrid approach for improving the design quality of web service interfaces. ACM Trans. Internet Technol. (TOIT) **19**(1), 1–24 (2018)
25. Palma, F., Moha, N., Tremblay, G., Gueheneuc, Y.G.: Specification and detection of SOA antipatterns in web services. In: Software Architecture, pp. 58–73 (2014)
26. Perepletchikov, M., Ryan, C., Frampton, K., Schmidt, H.: Formalising service-oriented design. J. Softw. **3**(2), 1–14 (2008)

27. Perepletchikov, M., Ryan, C., Tari, Z.: The impact of service cohesion on the analyzability of service-oriented software. IEEE Trans. Serv. Comput. **3**(2), 89–103 (2010)
28. Read, J., Pfahringer, B., Holmes, G., Frank, E.: Classifier chains for multi-label classification. Mach. Learn. **85**(3), 333 (2011)
29. Rodriguez, J.M., Crasso, M., Mateos, C., Zunino, A.: Best practices for describing, consuming, and discovering web services: a comprehensive toolset. Softw. Pract. Experience **43**(6), 613–639 (2013)
30. Rotem-Gal-Oz, A.: SOA Patterns. Manning Publications (2012)
31. Tsoumakas, G., Katakis, I.: Multi-label classification: an overview. Int. J. Data Warehous. Min. **3**(3), 1–13 (2007)
32. Wang, H., Kessentini, M., Ouni, A.: Bi-level identification of web service defects. In: International Conference on Service-Oriented Computing, pp. 352–368 (2016)
33. Wang, H., Kessentini, M., Ouni, A.: Interactive refactoring of web service interfaces using computational search. IEEE Trans. Serv. Comput. **3** 6–12 (2017)
34. Wang, H., Ouni, A., Kessentini, M., Maxim, B., Grosky, W.I.: Identification of web service refactoring opportunities as a multi-objective problem. In: IEEE International Conference on Web Services (ICWS), pp. 586–593 (2016)
35. Zhang, M.L., Zhou, Z.H.: Ml-knn: a lazy learning approach to multi-label learning. Pattern Recogn. **40**(7), 2038–2048 (2007)

A Contract Based User-Centric Computational Trust Towards E-Governance

Bin Hu[1,2]([✉]), Xiaofang Zhao[1], Cheng Zhang[1], Yan Jin[1], and Bo Wei[3]

[1] Institute of Computing Technology, Chinese Academy of Sciences, Beijing, China
`hubin@ncic.ac.cn`
[2] University of Chinese Academy of Sciences, Beijing, China
[3] The First Research Institute of the Ministry of Public Security, Beijing, China

Abstract. E-Government services are persistent targets of the organized crime by hackers, which hinders the delivery of services. Computational trust is an important technique for the security work of service providers (SPs). However, it relies on data collection about users' past behaviors conventionally from other SPs, which incurs the uncertainty of data and thus impacts the quality of data. Motivated by this issue, this paper proposes a novel smart contract based user-centric computational trust framework (UCCT) which collects the behavioral data of the user. It uses smart contract as a rational trustworthy agent to automatically monitor and manage the user's behaviors on the user side, so as to provide deterministic data quality assurance services for the computational trust. Furthermore, a privacy-preserving way of the data sharing is provided for the user and a personalized security mechanism for the SP. A new ledger is also introduced to provide a user-centric and efficient search. The results of experiments conducted on a Hyperledger Fabric based blockchain platform demonstrate that the time cost of user-centric ledger in UCCT can be less than 1 s. Moreover, even if a more complicated contract is provided, the improvement of transaction per second (TPS), which is made by UCCT, is not less than 8%.

Keywords: Computational trust · Data quality · Smart contract

1 Introduction

Domestic experiences show that the e-governance services are persistently targeted by hackers and organized crimes, which hinders the delivery of services and impact the confidentiality, integrity, and availability of information [1]. Services and users maintain a fragile trust relationship because of uncertainty and suspicion [2]. Computational trust is an appealing technique that provides a dynamic and measurable value to ensure flexible and controllable security in

Supported by the National Natural Science Foundation of China-Joint Fund for Basic Research of General Technology under Grant U1836111 and U1736106.

© Springer Nature Switzerland AG 2020
W.-S. Ku et al. (Eds.): ICWS 2020, LNCS 12406, pp. 133–149, 2020.
https://doi.org/10.1007/978-3-030-59618-7_9

e-Government services. To do that, it needs to collect past behaviors about the specific user from the other services. How to get a high quality behavioral data on e-Government ecosystem, however, is still challenging. Besides, little information is available on how internet-based data collection can be accomplished.

Computational trust in e-Government ecosystem does not have a centralized party that records users' past behaviors. The SP needs to communicate with other SPs whom the user has visited to collect the user's past behavioral data. Alexopoulos et al. [4] pointed out that the issues of trust, information responsibility, and consensus among nodes hindered the data collection process and brought uncertainty to the data quality. For example, Aguirre and Alonso attempted to share security alarms among different domains to enhance the global security [5] and Vasilomanolakis et al. transformed the data collection of intrusion detection into data sharing between different intrusion detection systems [6]. The trustworthiness between different nodes reduces confidence in data sharing [5,6]. Researches in [8,9,11,14] tried to collect data from the other services to correct the trust evaluation results. Yu et al. pointed out that the subjectivity [7] of the other services could not be ignored, especially when the other services are not interested in or may intentionally provide profitable information to maximize their own gain. More importantly, the collection of users' past behaviors conflicts with users' privacy protection demands. Atote et al. [10] pointed out some challenges in privacy protection of user information such as the sale of personal privacy data.

To address the above issues, this paper proposes UCCT, a new approach for the reliable computational trust towards the decentralized network. It uses a contract-based method to solve the problem of data quality uncertainty and privacy leaking. Users can deploy a smart contract to rationally supervise and manage their own interaction information, and control the data sharing in a privacy preserving way. Besides, the contract-based method provides a flexible and personalized security between the SP and the user.

The contributions of this paper are three-fold:

1. A smart contract based framework is proposed to supervise and manage users' past behaviors, which provides a high quality data and a privacy-preserving data sharing.
2. A user-centric ledger is proposed to provide an efficient search.
3. Experiments based on the blockchain platform are conducted to evaluate the reliability and the performance of UCCT.

The remainder of this paper is organized as follows: Section 2 reviews related work. Section 3 formulates the data quality uncertainty problem. Section 4 proposes the UCCT framework for data management and addresses the accompanying threat of data fraud resulting from local storage of behavior data. Section 5 demonstrates the experimental results. Section 6 concludes the paper.

2   Related Work

2.1   Trust Agent

It is common to build a trustworthy environment among people or machines. Jøsang [15] pointed out that instead of establishing a trust relationship between two perceptual entities, it was better to make a rational agent who acted as a proxy for perceptual entities to participate in the establishment of trust so that they would act only in accordance with rules and instructions. That means, only trust those users who have security agents [16]. Aberer et al. also stated that agents that store and process trust-related data could not be unconditionally trusted, and their malicious behaviors also need to be considered [3]. Hammadi et al. [20] introduced service-independent agents, which functioned as a third party, to perform trust evaluation that raised the single point failure [7].

2.2   Data Quality

Users' interaction data are distributed over different services, and it is difficult to collect the data of specified objects in an unreliable, decentralized network which greatly affected the data quality of computational trust. Aberer et al. [3] summarized three major factors that could affect the reliability of trust calculation results: the reliability of network, the trustworthiness of the other services that provide opinion, and the subjective uncertainty of the other services. Even so, Teacy et al. [8] proved that comprehensive direct and indirect interaction information are important to improve the quality of data and enhance the reliability of computational trust. For example, almost all of the trust studies, such as PET [11], TRAVOS [8], FIRE [9], Zhang-Cohen [12], ARICA [13], Dossier [14], integrated the indirect information for trust evaluation. The reliability of those indirect information, however, was uncertain. Besides, it is usually hard to get the indirect information. Furthermore, little information is available on how internet-based data collection can be accomplished.

2.3   Smart Contract

Contract is an important way of delivering service. Ruohomaa and Kutvonen pointed out that the explicit expression of implicit expectations through contracts can encourage more trust and reduce uncertainty [17]. However, the contract is usually closely connected with business. For example, Schnjakin et al. [18] presented the service and business in contract, giving users a clear service content. Due to the technical difficulty and business connection issues, the application of contract is challenging. The Smart Contract was first to the public at the presence of the blockchain platform named Ethereum [19], and it is widely adopted by many well-known blockchain platforms, such as the Hyperledger Fabric [21], to perform transactions. Smart contracts in blockchain can be enforced and the transactions can be traced, which provide a reliable decentralized environment. Many works have utilized the blockchain and the smart contract technology to solve their current security problems. For example, Chen et al. [22] deployed a cost-effective payment collection supervision system based on blockchain technology. With the use of smart contracts combined with the food industry standards, Tao et al. [23] performed an automatic food quality detection and warning of substandard food in the entire industrial chain. Yong et al. [24] addressed the problems of vaccine expiration and vaccine record fraud through the blockchain and the smart contract technology.

Our approach, named UCCT, uses the smart contract as a trustworthy rational agent to execute compiled, business independent computer code, monitor and manage user's past behavior data in a privacy-preserving way and provide high-quality data services for the computational trust. The contract takes the SP's trust model as an input and feeds back with evaluation results. Hence, the SP can flexibly adjust its model or parameters and build up a personalized security mechanism.

3   Problem Formulation

This section analyzes the problem studied in this research. Given a user, one SP needs to select a number of services from the neighbor nodes or the similar services to collect the past behavior data for the trust evaluation procedure since it does not know which services the user has accessed. It is difficult to gather all those data accurately. Due to the poor data quality, the computation trust cannot provide a reliable trust value.

Throughout the paper, the notations are listed as Table 1.

Table 1. Notations

Notation	Description
S	The service that can be provided to the user
C	The user that uses the service
SP_e	The SP that starts a trust evaluation about a C
SP_o	The SP that excludes the SP_e
D^c	The past behavior data collected about C
D_i^c	The past behavior data collected from service i about C
fact:SP_e& C	Direct information between SP_e and C
fact:SP_o& C	Direct information between SP_o and C
opin:SP_o& C	Indirect information between SP_o and C

As shown in Fig. 1, the evaluator SP (SP_e) needs two types of data from the given user (C, the assessed): the direct information (past behavior data between C and SP_e, denoted as fact:SP_e&C) and the indirect information (past behavior

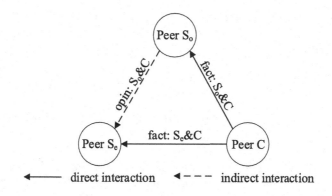

Fig. 1. Data sources of the trustworthiness calculation procedure

data between C and the other SPs (SP_o, those that C has been accessed exclude the SP_e). The SP_o, most of the time, can only provide the opinion on C, denoted it as opin:SP_o&C).

Assume that C has accessed n services, then the data D about C can be denoted as $D^c = \{D_i^c | i \in [1, n]\}$.

As illustrated in Eq. (1), SP_e tries to get the best high-quality data about the given C from the network. The ideal situation is that all these services provide the direct information.

$$arg\ max\ P(\text{the data quality of C}) = P\left(\bigcap_{i=1}^{n}(D_i^c = fact : S_i \& C)\right)$$

$$= \prod_{i=1}^{n} P(D_i^c = fact : S_i \& C) \tag{1}$$

But there are many factors affecting the data quality in a decentralized network. We define the three influencing factors mentioned in [3,4] as: N-network conditions, T-the credibility of the SP_o, and Q-the quality of the data provided by the SP_o. Then the key lies in how to improve the reliability of data quality and reduce the impact of N, T, and Q. For the convenience of analysis, we assume that N has only two status: reachable and unreachable. T is defined as providing data in a cooperative and non-cooperative way. Q is defined as providing the interaction record with C or other data opinions of non-client interaction records.

Assumption 1. *The network condition is reliable, that is, it will not affect SP_e to collect the data. However, the T status of SP_o and C is not confirmed, but if they get to cooperate, they will provide the data information about C's direct interaction behavior data.*

Then we can derive an Eq. (2).

$$P(D_i^c = fact : S_i \& C) = P(T_i = \text{cooperative}) \tag{2}$$

Assumption 2. *Both the network situation and the status of the other services* SP_o *are uncertain, but it can still be guaranteed that the direct information of C will be provided in the case of* SP_o*'s cooperation.*

It is noted that the network situation is independent of whether SP_o cooperates or not. Then further, we get the Eq. (3).

$$P(D_i^c = fact : S_i \& C) = P(N_i = \text{reachable}) \cdot P(T_i = \text{cooperative}) \qquad (3)$$

Assumption 3. *The network situation is uncertain, nor is the status of the other services* SP_o*, nor is there any guarantee that* SP_o *will provide information about C's direct interaction behavior even in case of* SP_o*'s cooperation.*

Similarly, we get the Eq. (4).

$$P(D_i^c = fact : S_i \& C) = P(N_i = \text{reachable}) \cdot P(T_i = \text{cooperative})$$
$$\cdot P(Q_i = \text{fact}) \qquad (4)$$

Combined with the Eq. (1) and (4), we will obtain the traditional method for data collecting on computational trust, which can be expressed as:

$$P(\text{the data quality of C}) = \prod_{i=1}^{n} [P(N_i = \text{reachable}) \cdot$$
$$P(T_i = \text{cooperative}) \cdot P(Q_i = \text{fact})] \qquad (5)$$

From the perspective of SP, the data quality is difficult to guarantee when there are multiple influencing factors and nodes involved. It can be noted that the Eq. (5) is an exponential function on nodes under the conditions of determined T, N, and Q factors. As the number of nodes increases, the probability of SP_e obtain a high-quality data of C drops down sharply. If the number of nodes is determined, it is a power function about probability. That is, the better network, the more cooperative and honest the SP_o, and the higher the data quality. However, the number of services accessed by users in Web Services is usually relatively large, which means that it might be difficult to ensure high data quality through this power function. From the perspective of the exponential function, we can obtain higher data quality only by reducing the number of participating nodes (the limit is 1). So, can we benefit from high data quality by reducing the number of participating nodes?

From the perspective of the C, it has all the behavior data that the SP needs. Hence, the Eq. (5) can be transformed into Eq. (6) and can quickly achieve a high probability of data quality if we can ensure that the C is honest and provides an accurate description about itself.

$$P(\text{the data quality of C}) = P(Q_c = \text{fact}) \qquad (6)$$

In that case, the influencing factors are constrained to C only, including the network between SP_e and C, the credibility of C, and the preservation of interaction history on C. Firstly, if the network between SP_e and C is poor, which means that SP_e cannot provide services to C, the computational trust of C is meaningless. Secondly, for C's credibility, this is the goal of SP_e for collecting data of C. As such, we only need to consider one factor, namely, how to promise the quality of data preserved on C. In other words, the problem of SP_e collecting a high-quality behavior data about C from the network is transformed into how C manages its data rationally and shares the data under privacy protection. Therefore, we need to design a good way for C to manage its data correctly and share its data securely in a decentralized environment.

4   User-Centric Based Computation Trust Framework

In this paper, we propose the smart contract based User-Centric Computation Trust framework (UCCT). As shown in Fig. 2, it uses the smart contract as a rational trustworthy agent to automatically monitor and manage the user's behaviors on the user side to provide deterministic data quality assurance services for the computational trust. Meanwhile, a different anti-fraud ledger is provided to avoid the data fraud. Besides, a personal module is provided for the SPs to publish its trust models or security policies here, and a trust certification (TC) profile is created for each user to organize the behavior data.

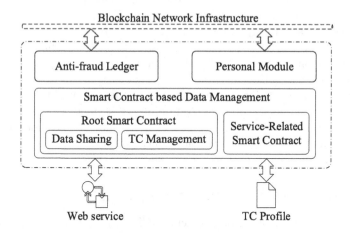

Fig. 2. UCCT framework

4.1   Trust Certification Profile

The accuracy of the past behavior data provided by the user can effectively reduce the uncertainty of data quality in computational trust. For the management of behavior data and unified data understanding, a behavior-management structure is needed. This article assigns a trust certification (TC) file for each user and the TC is automatically maintained and updated by the smart contract

```
<?xml version="1.0" encoding="UTF-8"?>
<T-certificate>
    <id>idvalue</id>
    <count>
        <total>totalcount</total>
        <benign>benigncount</benign>
        <malicious>maliciouscount</malicious>
    </count>
    <sequence>
        <behavior index=1>judgement</behavior>
        <!-- many more behavior -->
    </sequence>
    <hashCode>hashcodevalue</hashCode>
    <signature>signaturevalue</signature>
</T-certificate>
```

Fig. 3. TC structure for the user's past behavior data presented by XML schema

service. The TC structure shown in Fig. 3 includes a unique ID (*idvalue*), the total amount of interaction (*totalcount*), the number of good records (*benign-count*), the number of misbehavior records (*misbehaviorcount*), and the interaction sequence that including the interaction index (*index*), and the evaluation result (*judgement*). All those misbehaviors construct a Merkle tree which leads to a hash value (*hashcodevalue*) of this TC by a hash function, such as sha256. The signature (*signature*) is a security protection technique of cryptography.

The behaviors of the user are recorded in the TC file, and the disclosure of this file will cause harm to the user' privacy. For the consideration of the content security of the TC file and the limited storage on the blockchain, the TC file is stored locally on the user's disk and its data can only be operated after the smart contract is authorized by the user.

The sequence of behavior is quite important. Since older behaviors may lose influence quickly, majority of the outstanding trust models, taking [9,14] as examples, use the recency of the behaviors as a time weight function to give recent behaviors more weights than older ones. With the TC, A SP can easily learn much information on past behaviors about the user.

4.2 Smart Contract Based Data Management

Smart Contract based Data Management (SCDM) is responsible for the interaction recording on the user's side and sharing data in a privacy protection way for the SP and the user. All these works are done by managing the TC profile. Firstly, the data need to be updated into TC correctly and efficiently. Secondly, the validity of this TC profile needs to be checked before trust evaluation, including efficiency and correctness. Thirdly, the TC profile needs to be renewed simultaneously when the user is accessing multiple services at the same time.

To solve it, a two-level smart contract model is proposed for service. It includes a root smart contract (RSC) and a service-related smart contract (SRSC) management (SRSCM) module. The RSC is a critical part that is responsible for setting security policies, such as the privileges. It contains a TC management (TCM) module and a privacy-preserving based data sharing (DS) module which is responsible for TC updating and data sharing in a privacy preserving way respectively. The SRSC monitors interactions between this user and service, and the SRSCM manages all those SRSC, including SRSC creating and revoking.

When the user tries to access the SP, the following working flow is presented as a **S**mart contract based **T**rust **E**valuation **P**rotocol (STEP) in Fig. 4.

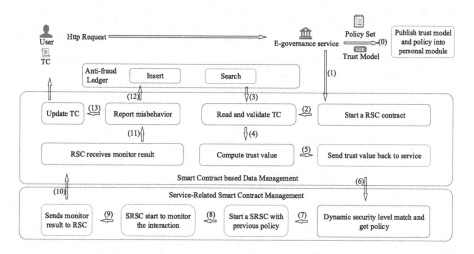

Fig. 4. UCCT procedure

1. The service starts a RSC contract.
2. The RSC's TCM reads and verifies the user's TC signature.
3. Anti-fraud checks to find out if there is a TC fraud.
4. The RSC's DS computes the trust value of this user by using the TC and the trust model in a privacy-preserving way.
5. The RSC sends this trust value back to the service.
6. The SRSCM uses this value to match a security level and to get an appropriate policy.
7. A SRSC is created by SRSCM to supervise the interactions between this user and SP under the previous policy.
8. The SRSC monitors interactions.
9. The monitor result is sent to RSC.
10. RSC receives result and find out if it is a misbehavior.
11. The misbehavior record is reported to the ledger first.
12. The ledger inserts this record into this user's data block.
13. RSC updates the TC.

The anti-fraud procedure is described in next subsection. The privacy preserving based data sharing is mainly realized by a smart contract. It takes the SP's trust model as an input and executes the computation on the user's side and feeds back the result to the SP. The SP does not get anything except the result.

4.3 Anti-fraud Ledger

The TC file, stores the user's behaviors on the user's side and gives the SP a full view of this user, can be accessed and manipulated easily. For example, at the time of t_1, the user behaves nicely and its credibility is high. Denote the TC at this time as TC_1 and suppose that this user keeps a copy of this TC_1. At the time of t_2, assume that, this user has performed some misbehaviors, and the trust value is significantly dropped down. Denote the TC at this time as TC_2. Then, in order to continue to maintain a good credibility, the user could use the old legitimate TC_1 to deceive the SP. We call it *TC fraud* and it can happen because the users can gain more resources or authority than they should have.

There would be two cases of *TC fraud*: 1) the user replaces the TC_1 with TC_2; 2) the user copies the TC_1 to a new machine to get a better service. Due to the advantages of blockchain ledger, it can properly fitted in to fix the problem. However, the current blockchain ledger is a chronological ledger, which means that transactions of all users are grouped in blocks in a chronological order and is time-consuming to get the data of a user. To solve this, a **De**centralized **b**lack **t**rust ledger (Debt) is brought in with user-centric feature in the chronological ledger. As illustrated in Fig. 5, it contains a Roster Router (RR) and several Roster Page Trees (RPT). Each RPT contains multiple Roster Page Nodes (RPN) which stands for different misbehavior users.

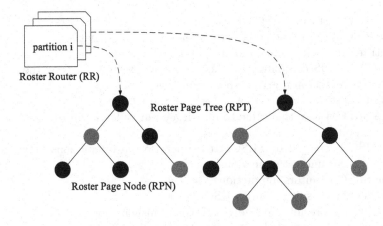

Fig. 5. Debt structure

The main step of anti-fraud check procedure is as follows:

1. The RR is a hash map which is used to locate a subledger-RPT.
2. The RPT is a red-black tree which gives an efficient search operation that is used to find out the position of the user's personal block-RPN.
3. A *RootHash* value is read from this RPN and compared with the *hashcodevalue* in TC to find out if they match.

All those operations mentioned above use the user's TC id as the key. The final goal is trying to get the *RootHash* value in RPN. As illustrated in the RPN structure in Fig. 6, the user's past misbehaviors are constructed in a Merkle tree and lead to a *RootHash* value stored in RPN head. Any new coming misbehavior will cause the change of *RootHash* value in the RPN head.

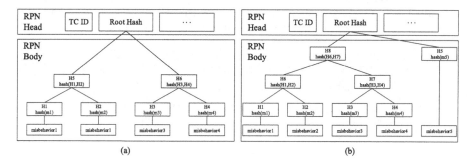

Fig. 6. RPN structure and Updating with a misbehavior record

4.4 Personal Module

Trust definition and trust model related thresholds are different among SPs. To provide a flexible security, SPs may define its own trust model or the other models here and set up with a changeable thresholds so as to adjust their security levels through users' trust performance. Besides, the management policies are also published here as the reference for the evaluation basis.

5 Performance Evaluation

This section experimentally evaluates UCCT through comparison with three different approaches from the viewpoint of cost performance.

5.1 Experiment Setup

Baseline. We build up a simple e-governance service by using GoAhead [25] and a Hyperledger fabric (HF) based platform. The deployment information of those two services is listed in Table 2. We define that the *index.html* contains two GET services: */edemo/type/get* and */edemo/get/list*. The first one requires a valid parameter *id* and should not be accessed frequently by the same user, and the second one must be accessed under authorization.

Table 2. Network deployment

Service	Deployment
Hyperledger Fabric (HF)	8 peers (192.168.0.11-18)
	3 order nodes (192.168.0.19-21)
	3 zookeepers (192.168.0.22-24)
	4 kafka (192.168.0.25-28)
E-governance Service	Deployed on 192.168.0.10
	/edemo/index.html, /edemo/type/select, /edemo/get/list

Comparing Approaches. The Debt and the SCDM are the most important parts in UCCT. We have implemented UCCT with a SCDM contract and a Debt ledger. For comparison, we have implemented three approaches:

- WS: the e-governance service as the baseline with no blockchain.
- WS(HF): the WS with the origial Hyperledger fabric platform and a simple transaction contract named chaincode_example02 [26].
- WS(HF-UCCT): the WS(HF) with a self defined SCDM contract named chaincode_rsc, which contains the procedure (1–5 and 11–14) described in Sect. 4.2 and the ledger replaced with Debt.

Evaluation Metric. We try to find out how much the UCCT would hinder the interaction separately and how much the UCCT would help with the misbehavior's detection. Three metrics are used: cost, transaction per second (TPS), and misbehavior detection efficiency (MDE).

Three series of experiments are designed. The first one focuses on the cost of insert and search operation in Debt with a batch number from 100 to 10,000,000. The second one focuses on cost and TPS, and tries to learn how much overhead the UCCT introduces to the service. The third one tries to find out the response speed to misbehaviors. We simulate a series of user behaviors and use the trust model defined in [9] to compute the trust value. Moreover, this value is used to setup the security level as shown in Table 3. At the initial state, a user has already executed 1,500 interactions with 30 misbehaviors. Then this user intentionally executes 20 misbehaviors to harm the SP through /edemo/type/get service with an illegal parameter value. The MDE compares and shows how fast those approaches could stop misbehaviors. With a bigger MDE, the service would detect the malicious user and end this interaction earlier.

Table 3 presents the corresponding settings. Each experiment is executed for 10 times and the results are averaged. All experiments are implemented in Go 1.13 and conducted on the same machine condition with Intel(R) Xeon(R) Bronze 3104 CPU @ 1.70 GHz and 64 GB memory, running Centos7 ×64.

Table 3. Experiment Settings

	Metric	Target	Test method
Experiment series	Cost	Debt	Batch size from 100 to 10,000,000
	TPS	WS	1 to 200 parallel service requests
		WS(HF)	
		WS(HF-UCCT)	
	MDE	WS	Static security policy
		WS(HF-UCCT)-α	Deny service if trust ≤ 60
		WS(HF-UCCT)-β	Deny service if trust ≤ 40

5.2   Results and Analysis

We introduce the UCCT to provide a reliable data collection method. There are two aspects of reliability. Firstly, *Data Quality Guarantee*. Collaboration data sharing between nodes is difficult. The UCCT provides a user-centric based data management through a rational smart contract service. Secondly, *Privacy Protection*. In this paper, the data are managed by the users themselves and shared without any raw data leaking. This makes sure that the behavior data are protected and accessed under authorization.

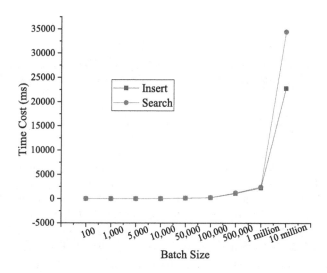

Fig. 7. Debt performance on search and insert operation

Figure 7 shows the overhead of Debt on insert and search operations. Both of them are tested through batch parameters. As demonstrated, when the batch size is within 500,000, the time cost is less than 1 s. When it increases to 1

million, the time cost is between 2.2 s to 2.4 s. This result gives us a hint on how
to set up the partition size of RR and the RPT sizes. The RPT size equals the
RPN number that it has, and if its maximum value is below 500,000, we can get
a high performance on ledger operation.

Experiment 2 compares the impact of smart contract which would hinder
the use of the framework. Figure 8 (a) gives a description of the average cost
of concurrent request and it is obvious that the smart contract based service
can cause performance degradation. To ease up this degradation, we set up an
evaluation cycle in practical application so that it would hinder the interaction
only at the beginning of several interactions. Figure 8 (b) shows the comparison of TPS between WS(HF) and WS(HF-UCCT), and we still make a TPS
improvement by 8%–10% under a more complicated contract in Fig. 4.

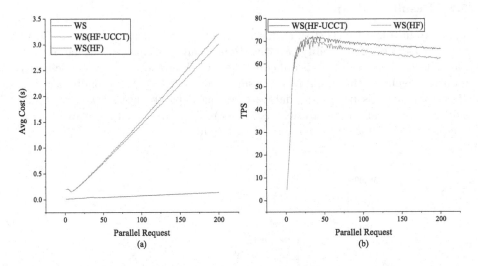

Fig. 8. UCCT overhead and TPS comparison

Figure 9 shows the results of experiment 3. By using a real trust model
defined in FIRE [9] and setting a connection between this value and security
level. The UCCT is capable of intercepting abnormal interactions flexibly. When
a user tried to access */edemo/type/select* or */edemo/get/list* in a wrong way for
many times. The WS is certainly capable of finding some of it by consuming
its resources, while it may run out of service. The UCCT, however, evaluate the
past behaviors and adjust the privilege dynamically, and is capable of protecting
itself. As shown in Fig. 9 (a), the WS(HF-UCCT)-α with a higher security level,
compared with the WS(HF-UCCT)-β, is capable of intercepting more misbehaviors. The WS, however, needs to process all those requests. In Fig. 9 (b), the
WS(HF-UCCT)-α intercepts 80% of the misbehaviors and the WS(HF-UCCT)-
β intercepts only 45%. That means, the SP can adjust those security parameters
flexibly to meet their own security demands.

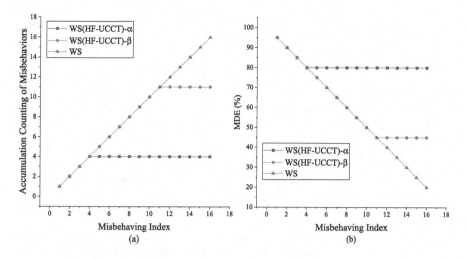

Fig. 9. Continuous misbehavior requests and MDE performance

6   Conclusion and Future Work

The data collection is fundamental to the computational trust. We propose the UCCT to provide SP with high-quality behavioral data of users and flexible and personalized security mechanism, and provide users with privacy protection based data sharing. The blockchain and smart contract technology, have no doubt, will significantly influence the way of dealing with trust related issues. Despite these benefits, the overhead is inevitable and the data management method based on blockchain requires delicate design. In addition, the ledger may need appropriate modifications to fit in the application. The ledger designed in this paper provides the capability of high-performance on insert and search operations. Even though, more work needs to be focused to provide a more reliable and efficient service. We plan to improve two parts of work, the smart contract based service performance and the Debt, which will provide a more efficient contract-based service and a more reliable user-centric ledger.

References

1. Kumar, B.S., Sridhar, V., Sudhindra, K.R.: A case study: risk rating methodology for E-governance application security risks. i-Manager's J. Softw. Eng. **13**(3), 39–44 (2019)
2. Grudzewski, W.M., Hejduk, I.K., Sankowska, A.: Trust management-the new way in the information society. J. Econ. Organ. Future Enterp. **2**(2), 2–8 (2008)
3. Aberer, K., Despotovic, Z.: Managing trust in a peer-2-peer information system. In: Proceedings of the Tenth International Conference on Information and Knowledge Management, pp. 310–317. ACM, Atlanta (2001)

4. Alexopoulos, N., Vasilomanolakis, E., Ivánkó, N. R., Mühlhäuser, M.: Towards blockchain-based collaborative intrusion detection systems. In: D'Agostino G., Scala, A. (eds.) CRITIS 2017, LNCS, vol. 10707, pp. 107–118. Springer, Cham (2017). https://doi.org/10.1007/978-3-319-99843-5_10
5. Aguirre, I., Alonso, S.: Improving the automation of security information management: a collaborative approach. IEEE Secur. Priv. **10**(1), 55–59 (2011)
6. Vasilomanolakis, E., Karuppayah, S., Mühlhäuser, M., Fischer, M.: Taxonomy and survey of collaborative intrusion detection. ACM Comput. Surv. (CSUR) **47**(4), 55 (2015)
7. Yu, J., Fan, W., Hong-Wu, Z., Li, D.: Survey on trust mechanisms in the environment of cloud computing. J. Chin. Comput. Syst. (2016)
8. Teacy, W. T., Patel, J., Jennings, N. R., Luck, M.: Coping with inaccurate reputation sources: experimental analysis of a probabilistic trust model. In: 4th International Joint Conference on Autonomous Agents and Multiagent Systems, pp. 997–1004. ACM, Utrecht (2005)
9. Huynh, T.D., Jennings, N.R., Shadbolt, N.R.: An integrated trust and reputation model for open multi-agent systems. Auton. Agent. Multi-Agent Syst. **13**(2), 119–154 (2006). https://doi.org/10.1007/s10458-005-6825-4
10. Atote, B. S., Zahoor, S., Dangra, B., Bedekar, M.: Personalization in user profiling: privacy and security issues. In: 2016 International Conference on Internet of Things and Applications (IOTA), pp. 415–417. IEEE, India (2016)
11. Liang, Z., Shi, W.: PET: a PErsonalized trust model with reputation and risk evaluation for P2P resource sharing. In: 38th Hawaii International Conference on System Sciences (HICSS-38 2005), pp. 201b–201b. IEEE, Big Island (2005)
12. Zhang, J., Cohen, R.: Evaluating the trustworthiness of advice about seller agents in e-marketplaces: a personalized approach. Electron. Commer. Res. Appl. **7**(3), 330–340 (2008)
13. Saeed, O., Shaikh, R.A.: A user-based trust model for cloud computing environment. Int. J. Adv. Comput. Sci. Appl. **9**(3), 337–346 (2018)
14. Botelho, V., Kredens, K. V., Martins, J. V., Ávila, B. C., Scalabrin, E. E.: Dossier: decentralized trust model towards a decentralized demand. In: IEEE 22nd International Conference on Computer Supported Cooperative Work in Design, pp. 371–376. IEEE, Nanjing (2018)
15. Jøsang, A.: The right type of trust for computer networks. In: Proceedings of the ACM New Security Paradigms Workshop. ACM (1996)
16. Samlinson, E., Usha, M.: User-centric trust based identity as a service for federated cloud environment. In: Fourth International Conference on Computing, pp. 1–5. IEEE, Tiruchengode (2014)
17. Ruohomaa, S., Kutvonen, L.: Trust management survey. In: Herrmann, P., Issarny, V., Shiu, S. (eds.) iTrust 2005. LNCS, vol. 3477, pp. 77–92. Springer, Heidelberg (2005). https://doi.org/10.1007/11429760_6
18. Schnjakin, M., Alnemr, R., Meinel, C.: Contract-based cloud architecture. In: Proceedings of the Second International Workshop on Cloud Data Management, pp. 33–40. ACM, Toronto (2009)
19. Wood, G.: Ethereum: a secure decentralised generalised transaction ledger. Ethereum Proj. Yellow Pap. **151**, 1–31 (2014)
20. Hammadi, A., Hussain, O.K., Dillon, T., Hussain, F.K.: A framework for SLA management in cloud computing for informed decision making. Cluster Comput. **16**(4), 961–977 (2012). https://doi.org/10.1007/s10586-012-0232-9
21. hyperledger fabric. https://www.hyperledger.org/projects/fabric. Accessed 20 Jan 2020

22. Chen, P.W., Jiang, B.S., Wang, C.H.: Blockchain-based payment collection supervision system using pervasive Bitcoin digital wallet. In: 2017 IEEE 13th International Conference on Wireless and Mobile Computing, Networking and Communications (WiMob), pp. 139–146 IEEE (2017)
23. Tao, Q., Cui, X., Huang, X., Leigh, A.M., Gu, H.: Food safety supervision system based on hierarchical multi-domain blockchain network. IEEE Access **7**, 51817–51826 (2019)
24. Yong, B., Shen, J., Liu, X., Li, F., Chen, H., Zhou, Q.: An intelligent blockchain-based system for safe vaccine supply and supervision. Int. J. Inf. Manag. **51**, 102024 (2019)
25. goahead. https://github.com/embedthis/goahead. Accessed 20 Jan 2020
26. chaincodeexample02. https://github.com/hyperledger/fabric/tree/v0.6/examples. Accessed 20 Jan 2020

Characteristics of Similar-Context Trending Hashtags in Twitter: A Case Study

Eiman Alothali[1], Kadhim Hayawi[2], and Hany Alashwal[1(✉)]

[1] College of Information Technology, United Arab Emirates University, Alain, UAE
halshwal@uaeu.ac.ae
[2] College of Technological Innovation, Zayed University, Abu Dhabi, UAE

Abstract. Twitter is a popular social networking platform that is widely used in discussing and spreading information on global events. Twitter trending hashtags have been one of the topics for researcher to study and analyze. Understanding the posting behavior patterns as the information flows increase by rapid events can help in predicting future events or detection manipulation. In this paper, we investigate similar-context trending hashtags to characterize general behavior of specific-trend and generic-trend within same context. We demonstrate an analysis to study and compare such trends based on spatial, temporal, content, and user activity. We found that the characteristics of similar-context trends can be used to predict future generic trends with analogous spatiotemporal, content, and user features. Our results show that more than 70% users participate in location-based hashtag belongs to the location of the hashtag. Generic trends aim to have more influence in users to participate than specific trends with geographical context. The retweet ratio in specific trends is higher than generic trends with more than 79%.

Keywords: Trend · Twitter · Spatiotemporal · Frequency · Context

1 Introduction

The popularity use of social networks services such as Facebook and Twitter by millions of users promote wealth of data. Social networks services, especially Twitter, have been a hub for many global events [1]. Understanding the posting behavior of participants of such networks can help in predicting pattern for future decisions. The use of hashtags has been known in social media arguments and exchange of thoughts in popular events. Hashtags are referred and constructed with only one specific word that starts with # symbol within the tweet [2]. The wealth of data of such hashtags has promoted studies from different domains to relate the impact of social media on real-life situation. Grover *et al.* investigated the social media discussions impact on voting behavior during election [3], and Gunaratne *et al.* studied temporal trends based on pro and anti vaccination discourse on Twitter [4]. Therefore, hashtags have been a target for adversaries attacks to manipulate its content by flooding it with unrelated content for hidden agenda [5, 6].

In Twitter, users can read or participate in popular hashtags (known as *trends*) during their evolving in time [7]. Such data can be used to display lists of recommendations.

© Springer Nature Switzerland AG 2020
W.-S. Ku et al. (Eds.): ICWS 2020, LNCS 12406, pp. 150–163, 2020.
https://doi.org/10.1007/978-3-030-59618-7_10

For example, the followers that a user follows are used by Twitter to recommend another list of followers based on his preferences [8]. Also, based on the user's location a list of popular hashtags within the preferred location is presented in trending list (user can change the location preference) [9]. However, an active trending hashtag can exceed the location of its people, if it has worldwide users participants or remained active in posting for hours or days. Such hashtag can be trending in worldwide list, as it evolves and remains active for certain time regardless of users' locations. Thus, it is important to study hashtag's features that can be a key start to explore users opinion, identify communities and influence, as well as to predict future outcome or detect adversaries' attacks.

One of the perspectives to study hashtags is to study them within a context. A similar-context hashtag can refer to specific topic or domain that the main keywords are linked somehow to each others [10]. For example, if we have a hashtags that discusses a topic directly or indirectly; such as #climate or #climatechange or #Globalwarming then, we can consider them all as similar-context hashtags as they address climate topic or domain. Therefore, a context trends refers to hashtags that are linked somehow as cause and effect perspective. For example: #bushfire is an effect of #climatechange and both are under the category of #climate.

Most of studies focus on content context from sentiment perspective [11–15]. Yaqub *et al.* analyzed Twitter content based on political context of 2016 US presidential elections to evaluate how the content represents public opinion accurately [14]. Another study presents information retrieval techniques models to identify key terms and context in Twitter [13]. Henry *et al.* proposed a data filter model for hashtags based on the context [16].

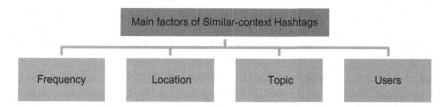

Fig. 1. Main factors of similar-context trending hashtags.

Understanding the spatiotemporal features of trending hashtags is essential to explore different behavior patterns of different trending events. As people interests are unlike, hashtags vary in their dynamics and context. Therefore, the main goal of this study is to investigate and discover part of spatial, temporal, content and user features in similar-context trends. We attempt to find how similar-context hashtags can support each other's or interact. We pose the following research questions:

RQ1: *What is the spatiotemporal behavior of similar-context trending hashtags?*
RQ2: *In case of location-based hashtags event, how often is participants belong to same location?*
RQ3: *What is the condition and ratio of tweets/retweet in similar-contexts hashtags?*

RQ4: *In similar-context trending hashtags, how often they share the same users?*

In summary, our findings indicate that spawning trending hashtags that are specific-trends induce generic trends. We found that more general trends attract more users from location of specific trends hashtags with similar-context with estimation of more than %70. Generic trends aim to have more influence in users to participate than specific trends with geographical context. We find specific-trends have more retweet than tweets, with more than 79%. The major contributions of this work are as follows:

- Conducted context-based study for similar trends according to selected criteria that help to better understanding of trending hashtags
- Studied growth of trends across time and location
- Characterize main factors that affect trending hashtags (Fig. 1): frequency, location, topic, and user activity.

The rest of the paper is organized as follows. Section 2 surveys the related work in the literature. Background and collection of data is presented in Sect. 3. Section 4 demonstrates our analysis of similar-context trending hashtags. Discussion of the results is in Sect. 5. Section 6 presents conclusion and future work.

2 Related Work

Studying different properties of social networks such as users interactions and information diffusion have been proposed in literature from the beginning of social networks. Many researches have discussed different aspects of Twitter's hashtags (Fig. 2). Guille *et al.* surveyed main study areas in the literature for information diffusion in Online Social Networks (OSNs) to three areas: to detect interesting topics, model diffusion process and identify influencers users [17]. Modeling OSNs information diffusion process can be either predictive modeling or explanatory modeling. In predictive modeling, the objective is to learn how the network graph has unfold by learning past diffusion traces of the network through temporal and spatial features. In explanatory modeling, the goal is to identify the complete cascade sequence to retrace the path of information diffusion. Therefore, diffusion modeling is helpful to understand information propagation in online social networks. Huang *et al.* proposed a predictive model framework using Deep-Neural-Network [18]. Their model aims to overcome the challenges of dynamicity and impact factors as hashtags change over time. Their solution is based on embedding dynamic and static factors and using a cumulative popularity value as a trigger.

Similarly, studies aimed to predict popularity of hashtags [19–21]. Xu *et al.* proposed a predictive model that is based on temporal analysis to estimate peak time and volume of bursting hashtags [19]. They found that tightness of diffusion network has a role in peak popularity of bursting hashtags. Some studies used differential equations in predicting information diffusion in social network [22, 23]. Davoudi *et al.* used linear ordinary differential equations (ODEs) to study temporal patterns as dynamic carrying capacity in order to predict influenced users [22]. Wang *et al.* used diffusive logistic (DL) equation to model temporal and spatial information diffusion [23]. Their model predicts the density of influence users on the early phase of information diffusion. The model achieved 92.08% accuracy in the first 6 h.

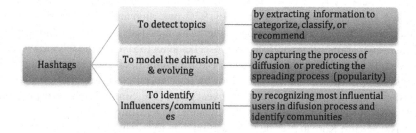

Fig. 2. Studying Hashtags in OSNs.

There are several studies to uderstand the temporal dynamics of information diffusion based on user influence in Twitter [22–25]. In a recent study, Stai *et al.* proposed an epidemic model for information spread in Twitter to model temporal behavior of a hashtag propagation [24]. The evaluation result of their model shows a constant infection rates for hashtags that belong to general topics or event-specific topics that have global impact for long duration of use. In general, they found that the type of hashtag has a role of the infection rate to increase or decrease over time.

Furthermore, another studies discuss the classification of trending topics in Twitter [26–30]. A recent study proposed a framework model called TORHID (Topic Relevant Hashtag Identification) that retrieves and identifies hashtags that are related to a specific topic in Twitter [26]. Their model starts with small tweets of a hashtag to work as seeds, then uses Support Vector Machine to classify new tweets as relevant or not. Their model achieved 67.25% accuracy. An analysis study for trending topics for the year of 2018 is proposed by [27]. They built their analysis on six criteria: lexical analysis, time to reach, trend reoccurrence, trending time, tweets count, and language analysis. Based on their studied dataset, they found that more than 17% of trending topics ranked for less than 10 min and more than 50% of studied trends couldn't hold for more than an hour. Lee *et al.* proposed a model to classify trending topics in Twitter into 18 categories such as sport, politics, etc. [28]. They used bag-of-words approach and network-based classification to build their model. Their model was able to classify trends with 65% and 70% accuracy. Another study proposed a model to classify trends into positive, negative, and neutral based on timestamp parameter [29]. The authors assume if the topic of a trend is promoting constructive idea then it is positive trend. The negative trend based on their model is a trend that aims to defame a person or organization while they consider a trend to be neutral if it is related to events such as sport or entertainment.

Hashtag recommendation based on topic model that use Latent Dirichlet Allocation (LDA) to retrieve similar information are proposed in [31–34]. Gupta *et al.* presented a large scale empirical analysis study of Twitter hashtags during the 2019 elections in India [31]. They explore the relationship between popularity of candidate on Twitter compared to the election outcome through analyzing positive and negative hashtags toward each candidate. The results show that influence score can predict the election outcome. Kim and Shim proposed a recommendation system called TWILITE [32]. They use a probabilistic modeling based on LDA, so their system recommends top-K users to follow and Top-K tweets to read. Godin *et al.* proposed unsupervised binary language classifier model based on Expectation-Maximization algorithm and Naïve bayes method [34]. Their model achieved 97.4% accuracy, 97% precision and 97.4% recall.

Understanding retweet dynamics in Twitter is another study area. Bi and Cho proposed a Bayesian nonparametric model to analyze user behavior [35]. Their model can determine automatically the parameter of the model based on the input data. Their model achieved for retweeting a topic a precision of 64%. A study by Ten *et al.* proposed a mathematical model to investigate and construct evolution of retweet graph [36]. Similarly, Ko *et al.* presented a mathematical model to measure Twitter dynamic shared-information in tendencies of public using South Korean presidential election data [37]. They measured information-shared based on two scales, hour and day. They concluded that by using day-scale, they could measure the public attention.

3 Background

In this section, we introduce background information about trends properties and our data collection.

3.1 Trends Properties

As mentioned earlier, in Twitter, trends are defined as popular hashtags that are selected based on Twitter's algorithms to decide popularity of any hashtag [7]. A hashtag is a tag symbol # attached to topic/keywords to group all related posts to make it easier for users to read and share written posts under topic tag [2]. As posting events are growing in a hashtag, and users involvement increases to some threshold, then a hashtag become trending based on rate of events (posts) per unit of time.

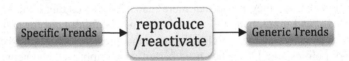

Fig. 3. Spawning Hashtags phenomenon in similar-context trends.

Hu *et al.* defines general terms to understand the three phases in popularity evolution prediction of hashtags [20]. These terms are burst, peak, and fade. Peak is the highest value that a trend reaches in a unit time. Burst is the increase rate before reaching peak of a trend. Fade is the opposite of burst, which indicates the decrease of posting and popularity to reach the inactive state.

In term of understanding trends features, Zubiaga *et al.* present a topology of trending topics based on its category in Twitter: news, ongoing events, memes, and communication [30]. They define 15 features for trending topics that are independent of language of tweets and focus on averages and diversity values that are extracted from tweets.

Similarly, Naaman *et al.* define five types of features for trending topics [7]. These features include content, interaction, time-based, participation, and social network features. Therefore, emerging trends can reveal information about global and local communities with geographic context.

3.1.1 Spawning Trends

Once a hashtag reaches popularity, it will remain in trends list for certain time that Twitter decides based on the count of tweets. Therefore, we observe hashtags that were created in certain incidents are reproduced and reactivated as a current incident happen again. For example, the first tweet using hashtag #cavefire was back to 2011 and the first tweet for #ClimateEmergency was back to 2009. We call this phenomenon as spawning trends (Fig. 3). It relates the specific trends with generic trends in similar-context. A specific trend is a trending hashtag that its keywords indicate a specific location or specific event or person. For example, #USA indicates a location based hashtag, and #USA2020elections indicates a specific event. While a generic trend is a trending hashtag that its keywords more general in term of scale to include wider participants regardless of location or any specific events and its usually to draw attention. Examples of generic hashtags are #climate, and #Football.

Table 1. Collected data statistics of each hashtag for the first 10 h

Hour	#CaveFire	#SydneyIsChoking	#ClimateEmergency	#ClimateAction
0	406	60	119	46
1	1165	1068	2111	276
2	852	1075	1588	231
3	426	825	1730	216
4	326	858	1469	353
5	300	855	1275	502
6	326	638	1233	711
7	422	263	1264	609
8	561	166	956	701
9	1058	133	932	672
10	810	77	784	778

3.2 Data Collection

In order to maintain validity of the model, we collect a real data from Twitter for the experiment. Our data collection process was performed using an available plug in tool called *Tweet Archiver* [38]. This tool collects tweets in present time. We used similar events hashtags for different dates and duration to capture for purpose of study and compare. For example, the hashtag of the incident of bushfire in USA "#CaveFire" tweets was collected for duration of nine days starting from the beginning of the event 25th Nov 2019 till 3rd December 2019. The total number of tweets and retweets that are collected is 15,377. The majority of this total took place in the first two days with 11,149 tweets/retweets.

A similar incident of bushfire was happening in Australia has lead to a hashtag "#SydneyIsChoking". Meanwhile, another similar-context hashtags where raised as well; #ClimateAction, and #ClimateEmergency. The last three hashtags data was collected in 10th –11th Dec 2019. The collection was conducted for 24 h and the total tweets/retweets is 35,961. Table 1 shows details information about collected data for the first 10 h for all hashtags. All trends were collected based on the trending list of each country and worldwide list.

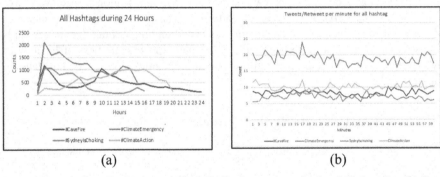

(a)   (b)

Fig. 4. Hashtags flow distribution (a) show distribution per hour for first 24 h duration. (b) Show distribution average per minute for each hashtag.

4   Similar-Context Hashtags Analysis

In this section, we demonstrate our analysis on the collected data of context-based trending hashtags. We focus on the four features: temporal, spatial, content, and user activity. We investigate each feature and observe the pattern based on arithmetic mean and frequency count.

4.1   Temporal Analysis

Twitter hashtags satisfied dynamic evolving distribution pattern. The trending hashtags as *events* are independent of each other. Therefore, the occurrence of one hashtag does not affect the probability of another hashtag to appear. Besides, each hashtag has an average rate of tweet/retweet per time period. The arrival of tweets/retweet in a hashtag is in a random pattern. Figure 4 shows the frequency of hashtags for first 24 h. It shows in (a) how a first hour has a sharp increase rate to reach peak in #Cavefire, #SydenyIsChoking, and ClimateEmergency. For #ClimateAction, it has a steady growth for more than 12 h then after the 15th hour, it starts to fade. This shows how generic trend is spawned from a similar-context specific trends.

In #ClimateEmergency and #SydneyIsChoking both remained inactive in 15th and 16th hours. For #CaveFire, it continues as this relates to more collection duration compared to other hashtags.

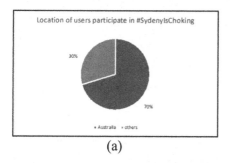
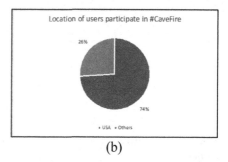

Fig. 5. (a) Specific-Trend percentage of users based on their location in #SydneyIsChoking. (b) Specific-Trend percentage of users based on their location in #CaveFire.

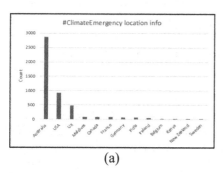
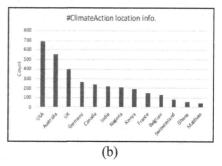

Fig. 6. (a) Generic-Trend participants location in #ClimateEmergency (b) Generic-Trend participants location in #ClimateAction

The information flow per minute as in Fig. 4 (b) shows almost similarity for #ClimateEmergency and #ClimateAction. This might indicate that both hashtags share similar users especially that the collection time for both hashtags was the same. In other words, shared users have included both hashtags in their tweets as shown in Table 2. This might also indicate that even if users are not similar in both hashtags, users put more than a hashtag in their posts as in Fig. 8.

4.2 Spatial Analysis

We analyze our data spatial information to identify the location of users participated among all context-based trends in our datasets. We preprocess our data and retrieve location information to verify users participation based on their given location. We found that *specific-trend* hashtags #SydneyIsChoking and #CaveFire shows in Fig. 5 (a) and (b) have a percentage of 70% and 74% in respectively for users who belong to location of these hashtags. It is good to note that both specific-trends data was collected in different dates.

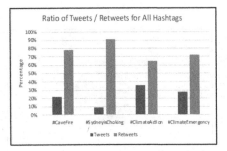
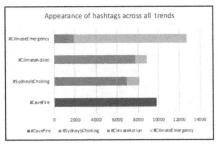

Fig. 7. Ratio of tweets and retweets in studied hashtags.

Fig. 8. Appearance of hashtags keywords across the studied hashtags.

On the other hand, the *generic-trend* hashtags #ClimateAction and #ClimateEmergency as shown in Fig. 6 (a) and (b) have more participants from specific-trends hashtag locations: Australia, and USA in the first two places. However, users from Australia have the largest share with a percentage of 60% in #ClimateEmergency and 21% in #ClimateAction. While users from USA have a total of 19%, 17% in respectively.

4.3 Content

Hashtag is a mixture of original tweets or retweets of other authors. As shown in Fig. 7, it is very clear that retweets are more than tweets no matter how long the hashtag remains. This is applied for all generic and specific trends. However, the retweet in specific trends is above 78% while in generic trends is below 72%. This indicates that specific trends users are reporting of incident more than discussing it. In terms of original tweets, hashtags with generic trends have about 31% while in specific it is 18%.

In terms of appearance of each hashtag across the whole data set content, Fig. 8 shows that the majority of appearance for each hashtags is about 16% for #SydneyIsChoking, #climateAction, and 17% #ClimateEmergency. The case of #CaveFire to not appear in other hashtags is due to the timing of collection data as the incident trend was previous to the rest. However, within its data there are 58 appearances for #ClimateEmergency but not vise versa. This emphasizes the spawning trend phenomenon, where the sequence of events impacts the popularity and reproduce/reactivate hashtag in similar-context. Besides, Fig. 8 shows that using same hashtags within same tweets is minimal in general with less than 17%.

4.4 Users Activity

In order to analyze users activity, we measure the activity of users compared to the content. In other words, we explore the participation of users per tweet/retweet as in

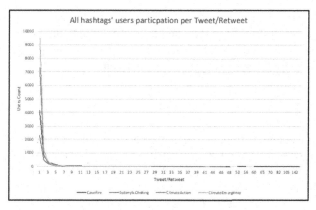

Fig. 9. The participation of users based on their tweet/retweet across all studied hashtags

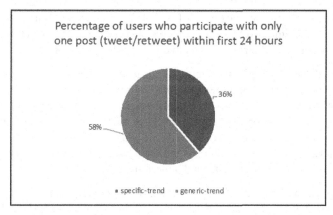

Fig. 10. Percentage of users who participate with only one tweet/retweet for studied sample of specific-trends and generic-trends

Fig. 9 and Fig. 10. The majority of users have post only one tweet/retweet across the studied hashtags as Fig. 9. Users who have been recorded with the highest activity was in #CaveFire, one account with 285 posts found to belong to a climate news account. In Fig. 10, generic trends' users tend to participate with 58% with one post.

Table 2. Users appearance a cross other hashtags.

	#CaveFire	#SydneyIsChoking	#ClimateAction	#ClimateEmergency
#CaveFire		22	88	127
#SydneyIsChoking			**545**	**1710**
#ClimateAction				**1864**
#ClimateEmergency				

In Table 2, the duplicate accounts that found across hashtags shows that generic trends have more duplicates of users. Therefore, it can imply the influence of generic trends compare to specific trends. In other words, it shows how specific trends events can lead to re-activate the generic trends in similar-context.

5 Discussion

The results of our analysis provide a strong indication that we can use the characteristics of similar-context specific trends to predict similar future generic trends in spatiotemporal, content, and user features. The findings show that users who participate with previous local hashtag within the same context tend to participate in future hashtags within context. We summarize our findings in the following:

Information Flow (RQ1): The dynamics of trends tend to change based on the popularity of topics and influence of users. Myers *et al.* found that 71% of influence in Twitter diffusion volume is due to internal events [39]. Therefore, similar-context hashtags that are trending in same time have evolution influence on each other, as users tend to include more similar hashtags in their posts as shown in Fig. 4b and Fig. 8.

Local vs Global Trends (RQ 2): In specific-trends it shows that users belong to the event location, participate more in their geographical trends. This supports a study [40] that indicates that popular topics across regional boundaries while specific topics remain within their region. Therefore, specific-trends have more shares of its location users as a hashtag gets popular. This can be justified by comparing Australia's users to USA's, where both have incidents of bushfire. As the incident was going on in Australia, it had more participants in both the specific and generic-trends.

Content (RQ 3): We find specific-trends have more retweet than tweets, with more than 79%. This can be explained, as within a specific location people tend to spread the information more than to create it. This is due to the fact of dealing with specific incident trends as news [30]. Results show that the percentage of tweets within generic trends is approximately half of the retweet percentage of its trends. This indicates the discussion dialogs and comments compared to specific trends. Also, the appearance of each hashtag across other hashtags is about 16% for majority of generic trends. This can be an indication of different users who participate within one of the studied hashtags and trying to promote other similar-context hashtags as well.

User activity (RQ 4): In general, the results show that users participate with majority with one post (tweet or retweet) within the first 24 h of a trending hashtag. Generic trends aim to have more participants with one post with 58% compared to specific trends with only 36% for one post per user. This indicates The Spawning phenomena where users tend to reactivate generic trends from specific trends considering duration and timing of popularity. Therefore, generic trends gain more shared users with specific trend in the same context.

6 Conclusion and Future Work

In this paper, we investigate similar-context trending hashtags in Twitter. Our analysis shows that participants in specific-trend related to location based hashtags represents more than 70% of participants. Participants in specific-trends have less user's activity in terms of creating tweets compared to generic trends in the first 24 h. The ratio of tweet and retweet shows that generic trends aim to have more discussions between users, while in specific trends users aim to retweet more. Based on our data, generic trends found to share more users with other specific and generic trends in general with a percentage of 16%.

For future work we might explore other associate features with more dataset sample to detect communities with similar-context trends. Also, we might compare the different context based hashtags with each other to find how they might change in terms of domain and users, for example political trends vs health.

References

1. Kwak, H., Lee, C., Park, H., Moon, S.: What is Twitter, a social network or a news media? In: Proceedings of the 19th International Conference on World Wide Web, pp. 591–600 (2010)
2. Ma, Z., Sun, A., Yuan, Q., Cong, G.: Tagging your tweets: a probabilistic modeling of hashtag annotation in twitter In: Proceedings of the 23rd ACM International Conference on Conference on Information and Knowledge Management, pp. 999–1008 (2014)
3. Grover, P., Kar, A.K., Dwivedi, Y.K., Janssen, M.: Polarization and acculturation in US Election 2016 outcomes – Can twitter analytics predict changes in voting preferences. Technol. Forecast. Soc. Change **145**, 438–460 (2019). https://doi.org/10.1016/j.techfore.2018.09.009
4. Gunaratne, K., Coomes, E.A., Haghbayan, H.: Temporal trends in anti-vaccine discourse on Twitter. Vaccine **37**(35), 4867–4871 (2019)
5. Zhang, Y., Ruan, X., Wang, H., Wang, H., He, S.: Twitter trends manipulation: a first look inside the security of twitter trending. IEEE Trans. Inf. Forensics Secur. **12**(1), 144–156 (2016)
6. Alothali, E., Zaki, N., Mohamed, E.A., Alashwal, H.: Detecting social bots on Twitter: a literature review. In: 2018 International Conference on Innovations in Information Technology (IIT), pp. 175–180 (2018)
7. Naaman, M., Becker, H., Gravano, L.: Hip and trendy: Characterizing emerging trends on Twitter. J. Am. Soc. Inf. Sci. Technol. **62**(5), 902–918 (2011)
8. Twitter. How to receive recommendations from Twitter (2020). https://help.twitter.com/en/managing-your-account/how-to-receive-twitter-recommendations. Accessed 02 Jan 2020
9. Twitter. Twitter trends FAQs (2020). https://help.twitter.com/en/using-twitter/twitter-trending-faqs. Accessed 02 Jan 2020
10. Tan, Y., Shi, Y., Tang, Q.: Data mining and big data. In: Third International Conference, DMBD 2018, Shanghai, China, 17–22 June 2018, Proceedings, vol. 10943. Springer (2018). https://doi.org/10.1007/978-3-319-93803-5
11. Aisopos, F., Papadakis, G., Tserpes, K., Varvarigou, T.: Content vs. context for sentiment analysis: a comparative analysis over microblogs. In: Proceedings of the 23rd ACM Conference on Hypertext and Social Media, pp. 187–196 (2012)
12. Vanzo, A., Croce, D., Basili, R.: A context-based model for sentiment analysis in twitter. In: Proceedings of Coling 2014, The 25th International Conference on Computational Linguistics: Technical papers, pp. 2345–2354 (2014)

13. Katz, G., Ofek, N., Shapira, B.: ConSent: context-based sentiment analysis. Knowl. Based Syst. **84**, 162–178 (2015)
14. Yaqub, U., Chun, S.A., Atluri, V., Vaidya, J.: Analysis of political discourse on twitter in the context of the 2016 US presidential elections. Gov. Inf. Q. **34**(4), 613–626 (2017)
15. Yadav, P., Pandya, D.: SentiReview: sentiment analysis based on text and emoticons. In 2017 International Conference on Innovative Mechanisms for Industry Applications (ICIMIA), pp. 467–472 (2017)
16. Henry, D., Stattner, E., Collard, M.: Filter hashtag context through an original data cleaning method. Procedia Comput. Sci. **130**, 464–471 (2018)
17. Guille, A., Hacid, H., Favre, C., Zighed, D.A.: Information diffusion in online social networks: a survey. ACM Sigmod Rec. **42**(2), 17–28 (2013)
18. Huang, J., Tang, Y., Hu, Y., Li, J., Hu, C.: Predicting the active period of popularity evolution: a case study on Twitter hashtags. Inf. Sci. (Ny) **512**, 315–326 (2020). https://doi.org/10.1016/j.ins.2019.04.028
19. Xu, W., Shi, P., Huang, J., Liu, F.: Understanding and predicting the peak popularity of bursting hashtags. J. Comput. Sci. **28**, 328–335 (2018). https://doi.org/10.1016/j.jocs.2017.10.017
20. Hu, Y., Hu, C., Fu, S., Fang, M., Xu, W.: Predicting key events in the popularity evolution of online information. PLoS ONE **12**(1), e0168749 (2017)
21. Kong, S., Mei, Q., Feng, L., Ye, F., Zhao, Z.: Predicting bursts and popularity of hashtags in real-time. In: Proceedings of the 37th International ACM SIGIR Conference on Research & Development in Information Retrieval, pp. 927–930 (2014)
22. Davoudi, A., Chatterjee, M.: Prediction of information diffusion in social networks using dynamic carrying capacity. In: 2016 IEEE International Conference on Big Data (Big Data), pp. 2466–2469 (2016)
23. Wang, F., Wang, H., Xu, K.:Diffusive logistic model towards predicting information diffusion in online social networks. In: 2012 32nd International Conference on Distributed Computing Systems Workshops, pp. 133–139 (2012)
24. Stai, E., Milaiou, E., Karyotis, V., Papavassiliou, S.: Temporal dynamics of information diffusion in twitter: modeling and experimentation. IEEE Trans. Comput. Soc. Syst. **5**(1), 256–264 (2018)
25. Raghavan, V., Ver Steeg, G., Galstyan, A., Tartakovsky, A.G.: Modeling temporal activity patterns in dynamic social networks. IEEE Trans. Comput. Soc. Syst. **1**(1), 89–107 (2014)
26. Figueiredo, F., Jorge, A.: Identifying topic relevant hashtags in Twitter streams. Inf. Sci. (Ny) **505**, 65–83 (2019)
27. Annamoradnejad, I., Habibi, J.: A comprehensive analysis of twitter trending topics. In: 2019 5th International Conference on Web Research (ICWR), pp. 22–27 (2019)
28. Lee, K., Palsetia, D., Narayanan, R., Patwary, M.M.A., Agrawal, A., Choudhary, A.: Twitter trending topic classification. In 2011 IEEE 11th International Conference on Data Mining Workshops, pp. 251–258 (2011)
29. Saquib, S., Ali, R.: Understanding dynamics of trending topics in Twitter. In: 2017 International Conference on Computing, Communication and Automation (ICCCA), pp. 98–103 (2017)
30. Zubiaga, A., Spina, D., Fresno, V.: Real-time classification of twitter trends. J. Assoc. Inf. Sci. Technol. **66**(3), 462–473 (2015)
31. Gupta, S., Singh, A.K., Buduru, A.B., Kumaraguru, P.: Hashtags are (not) judgemental: the untold story of Lok Sabha elections 2019 (2019). arXiv Prepr. arXiv:1909.07151
32. Kim, Y., Shim, K.: TWILITE: a recommendation system for Twitter using a probabilistic model based on latent Dirichlet allocation. Inf. Syst. **42**, 59–77 (2014)
33. She, J., Chen, L.: Tomoha: topic model-based hashtag recommendation on twitter. In: Proceedings of the 23rd International Conference on World Wide Web, pp. 371–372 (2014)

34. Godin, F., Slavkovikj, V., De Neve, W., Schrauwen, B., de Walle, R.: Using topic models for twitter hashtag recommendation. In: Proceedings of the 22nd International Conference on World Wide Web, pp. 593–596 (2013)

35. Bi, B., Cho, J.: Modeling a retweet network via an adaptive bayesian approach. In: Proceedings of the 25th International Conference on World Wide Web, pp. 459–469 (2016)

36. ten Thij, M., Ouboter, T., Worm, D., Litvak, N., van den Berg, H., Bhulai, S.: Modelling of trends in twitter using retweet graph dynamics. In: Bonato, A., Graham, F.C., Prałat, P. (eds.) WAW 2014. LNCS, vol. 8882, pp. 132–147. Springer, Cham (2014). https://doi.org/10.1007/978-3-319-13123-8_11

37. Ko, J., Kwon, H.W., Kim, H.S., Lee, K., Choi, M.Y.: Model for Twitter dynamics: public attention and time series of tweeting. Phys. A Stat. Mech. its Appl. **404**, 142–149 (2014)

38. Agarwal, A.: Tweet Archiver. https://gsuite.google.com/. Accessed 20 Oct 2019

39. Myers, S.A., Zhu, C., Leskovec, J.: Information diffusion and external influence in networks. In: Proceedings of the 18th ACM SIGKDD International Conference on Knowledge Discovery and Data Mining, pp. 33–41 (2012)

40. Ardon, S., et al.: Spatio-temporal analysis of topic popularity in twitte (2011). arXiv Prepr. arXiv:1111.2904

Finding Performance Patterns from Logs
with High Confidence

Joshua Kimball[✉], Rodrigo Alves Lima, and Calton Pu

Georgia Institute of Technology, Atlanta, GA 30332, USA
{jmkimball,ral,calton.pu}@gatech.edu

Abstract. Performance logs contain rich information about a system's state. Large-scale web service infrastructures deployed in the cloud are notoriously difficult to troubleshoot, especially performance bugs. Detecting, isolating and diagnosing fine-grained performance anomalies requires integrating system performance measures across space and time. To achieve scale, we present our *megatables* approach, which automatically interprets performance log data and outputs millibottleneck predictions along with supporting visualizations. We evaluate our method with three illustrative scenarios, and we assess its predictive ability. We also evaluate its ability to extract meaningful information from many log samples drawn from the wild.

Keywords: Cloud computing · Performance debugging · Anomaly detection · Data mining · Log data analysis · Information integration

1 Introduction

Cloud applications—especially those subjected to "bursty" workloads like ecommerce or social networking platforms—confront a common performance bug termed the "long-tail latency problem." This pathology is characterized by a small number of requests taking seconds to return even though most requests only take a few milliseconds to complete. Businesses have reported the economic impact this problem poses. Amazon found that every increase of 100 ms in page loading time is correlated to roughly 1% loss in sales; similarly, Google found that a 500 ms additional delay to return search results could reduce revenues by up to 20% [1, 2].

This is a puzzling problem because requests with long response times (order of seconds) start to happen at low CPU utilization levels (around 40%), when none of the hardware resources is anywhere near saturation on average. According to the millibottleneck theory of performance bugs, transient resource bottlenecks on the order of milliseconds, called millibottlenecks, can propagate through a distributed system via RPC calls and have their effects amplified, causing severe performance bugs despite their very short lifespan [3]. Given the evolution of cloud-enabled microservice architectures and their inherent reliance on RPC calls, performance has become less predictable. Moreover, performance anomalies have become harder to diagnose given the number of independent services and potential dependencies among them—possibly exponential in the number of services comprising such a system.

© Springer Nature Switzerland AG 2020
W.-S. Ku et al. (Eds.): ICWS 2020, LNCS 12406, pp. 164–178, 2020.
https://doi.org/10.1007/978-3-030-59618-7_11

Previous Work. Recent approaches like DeepLog operate over arbitrary text and attempt to isolate "macro-level" system events like crashes [4]. Seer relies on distributed event traces and microbenchmarks to mine microservice QoS violations [5]. Our automated approach operates over the diverse performance monitoring outputs with the objective of isolating much more precise (shorter and transient) performance anomalies.

In this paper, we show our system for automatically extracting data from performance monitoring log files to identify millisecond-scale performance pathologies—*megatables*. Our method goes beyond extraction as our approach interprets and analyzes the signals or performance patterns inherent in performance data and isolates those interesting performance phenomena.

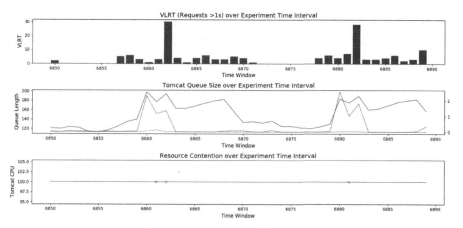

Fig. 1. Tomcat JVM Millibottleneck

Example 1. Figure 1 is a representative set of graphs necessary for effectively isolating and diagnosing millibottlenecks. In this case, these graphs correspond to a millibottleneck induced by JVM garbage collection. Each graph corresponds to a specific diagnostic step. The top graph shows the number of requests associated with very long response time, defined as requests exceeding 1 s to be processed, for each 50 ms interval. The middle graph depicts the queue size of each component for each 50 ms window. We determine the size of a queue using the number of requests waiting to be processed by the given component during each interval. The bottom graph depicts the resources that are temporarily saturating over the same interval. In this case, the Tomcat node's CPU is saturated due to Java Garbage Collection. This period of saturation coincides with the appearance of VLRT and the growth in queue size among dependent components. Given the correlation among these three variables—number of VLRT, component queue size and resource utilization—we can conclude the Tomcat CPU millibottleneck is induced by the Tomcat node's JVM Garbage Collection process. We have detailed this millibottleneck and its diagnostic procedure, briefly explained here, in our prior work [6]. *megatables* automates this process.

Challenges. Detecting and isolating millibottlenecks is particularly challenging due to their very short lifespan. According to the Sampling Theorem, multi-second sampling periods are insufficient to detect sub-second phenomena. On the other hand, an increased sampling frequency suggests increased logging overhead. Finding the right balance between these competing objectives is non-trivial. Moreover, understanding how millibottlenecks propagate across a system requires logging inter-node communication such that the message latency can be measured at fine-grained timescales, which most native logs do not contain. Extracting relevant resource data from a variety of performance monitor logs requires a flexible approach to data extraction. Thirdly, systematically identifying millibottlenecks requires the careful extraction and integration of important event and resource data spanning an entire system topology, i.e. space and time. Due to the diversity of millibottlenecks, a system needs to be able to make inferences from the performance patterns inherent in the data. It is impossible to know *a priori* which features are relevant to a given situation. Moreover, a system needs to be able to generalize across different types of millibottlenecks.

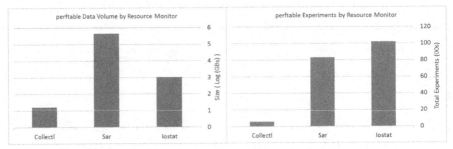

Fig. 2. Megatables Usage. Left graph shows the amount of experiment data extracted by megatables. Right graph shows the number of experiments where megatables was used.

Contributions. *megatables* begins by extracting data from component and resource monitor logs. Next, we transform the data into features salient to isolating and detecting millibottlenecks. We feed these features into machine learning models. Specifically, we leverage a Teamed Classifier approach. In this design, models are trained for each type of system and millibottleneck. Each model outputs one classification decision, and the decision with the highest confidence becomes the decision of the team. Finally, new data is fed into our trained models to generate predictions and supporting visualizations automatically.

To isolate and diagnose nuanced, fine-grained performance anomalies, we need to support a broad array of experimental configurations, since these bugs can materialize under a range of conditions. As Fig. 2 shows, we have run over 20,000 experiments generating over 100 TB of data spread across 400K various log files generated by our experimental computer science infrastructure, *elba*. As our first contribution, we present three illustrative examples of megatables ability to automatically diagnose millibottlenecks associated with different resources. Secondly, we demonstrate megatables models' coverage and accuracy. Figure 2 shows megatables data extraction covers 98% of the performance data generated by our infrastructure. As our final contribution, we show megatables coverage of performance data from the wild. Specifically, we gather a large random sample of performance log data from the wild, i.e. Github, and use megatables to process it.

2 Definitions

Recently, researchers have found that millibottlenecks, also termed very short bottlenecks (VSBs) or transient bottlenecks in the literature, can cause very long response time requests (VLRTs), which are those that take one to two orders of magnitude longer to complete than average [6, 7].

Definition 2.1 (Millibottlenecks). Very short-lived resource saturations caused by an underlying resource contention. They are associated with inducing very long response time requests through an intermediary transmission mechanism termed cross-tier queue overflow.

Definition 2.2 (Very Long Response Time Requests). Very long response time requests (VLRT) are those requests that exceed some critical threshold to be returned, for example requests associated with response times exceeding 1 s. VLRT requests are often masked by the normal requests that only take a few milliseconds, particularly when the response time is averaged over (typical) measurement periods of minutes.

Millibottlenecks and their associated VLRT requests appear and disappear on the order of hundreds of milliseconds. For example, Fig. 1 shows request latency over two short intervals where the number of VLRT requests during each 50 ms window exceeds 6x the average, i.e. the two large peaks.

Definition 2.3 (Point-in-Time Response Time). It is the average system response time over a defined interval using the time it takes requests initiated within the interval to complete a round trip.

VLRT requests can occur for very different reasons. Potential root causes span different system levels, including CPU dynamic voltage and frequency scaling (DVFS) control at the architectural layer [8], Java garbage collection (GC) at the system software layer [6], virtual machine (VM) consolidation at the VM layer [9], and performance interference of memory thrashing [10].

3 Illustrative Scenarios

In this section, we discuss how we used megatables to automatically isolate and detect two millibottlenecks due to different sources of resource contention. The first is a bottleneck induced by database (Mysql) disk IO. The other bottleneck is induced by dirty pages being flushed to disk. megatables automatically generates the following graphs at the end of its processing.

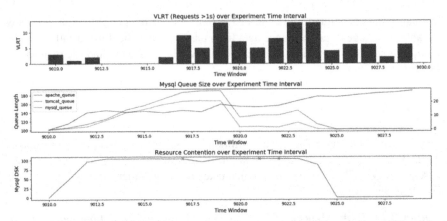

Fig. 3. Database IO as Millibottleneck. These provide visual evidence of a millibottleneck indicated by the appearance of VLRT, queue extension and temporary resource saturation.

3.1 Database IO as Millibottleneck

In Fig. 3, we illustrate megatables ability to isolate a millibottleneck due to Mysql's temporary disk saturation. Specifically, we review the period where the number of VLRT requests begins to grow and remains above 5. We can see the number of VLRT requests begins to decline quickly eventually returning to 0 in less than 250 ms.

To understand what occurs during this interval, *megatables* begins by extracting request traces generated by our specialized event tracing framework, milliScope, found in component logs. This data captures execution flow dependencies. As the middle figure shows, we observe obvious Cross-Tier Queue Overflow evidenced by the components' queue lengths elongating over the period of interest. Megatables uses the extracted request trace data to calculate a few metrics every 50 ms: point-in-time response time, the number of VLRT requests, i.e. those exceeding 1 s, and component queue length.

Megatables also extracts resource data from performance logs output by resource monitors. This data provides a representation of system state. In our scenario, there was one resource monitor, *collectl,* measuring CPU, Disk Memory and Network at 50 ms intervals. We see Mysql's disk is temporarily saturated, i.e. utilization reaches 100%, but returns to 0% after approximately 300 ms from the first moment it saturates. Megatables represents the extracted data for each resource category as a multivariate timeseries.

As mentioned earlier, megatables uses a data-driven approach to detect and analyze millibottlenecks. In short, it learns state and event-specific patterns consistent with the presence of millibottlenecks by relying on machine learning to systematically identify such patterns. In our case, diagnosing this database IO millibottleneck requires finding patterns where events such as the number of VLRT, the number of queued requests and the average Point-in-Time response time are maximal at the same time as state indicators such as Mysql disk resources are temporarily saturated.

We transform the event-based metrics and system state data into salient numerical features for detecting millibottlenecks. Specifically, we apply fixed-width windows to the event-based metrics mentioned earlier like point-in-time response time, the number of VLRT requests, i.e. those exceeding 1 s, and component queue length to create feature column vectors. These feature column vectors are concatenated into a matrix of row vectors such that each row represents a sample. In this scenario, we would construct an event feature column vector as follows: {PIT, VLRT, Apache_Queue, Tomcat_Queue, Mysql_Queue}.

We construct system state column feature vectors from the extracted resource data in a similar fashion. Each resource measurement every 50 ms is a component of a fixed width vector. In this case, each components' CPU, Disk, Memory and Network utilization are vector components. In our scenario, we would construct a state column vector as follows: {Apache_CPU, Apache_Disk, Apache_Mem_Used, Apache_Net_Bandwth, ...}. Like the event vectors, these vectors are concatenated into a matrix of row vectors where each row of the matrix is a sample.

We model the problem of determining the existence of a millibottleneck over some interval of time as a multi-class classification problem. To identify millibottlenecks, we use models trained over previously labeled data. Our labels indicate whether a millibottleneck is present, and if one exists what kind it is. These models are used to predict labels for each event and state matrix sample. In this case, the model indicates the presence of a Mysql (database) IO millibottleneck indicated by the red "X's" in the bottom figure.

3.2 Memory Dirty Page as Millibottleneck

In Fig. 4, we illustrate megatables ability to isolate a millibottleneck due to memory dirty page being flushed to disk. Specifically, we review the period where the number of VLRT requests begins to grow. We can see the number of VLRT requests begins to decline quickly eventually returning to 0 in less than 250 ms.

As in the prior situation, *megatables* begins by extracting request traces and calculates the associated event metrics. As the middle figure shows, we observe obvious Cross-Tier Queue Overflow evidenced by the components' queue lengths elongating over the period of interest. As before, megatables also extracts all resource data from the pertinent performance logs, in this case, *collectl*. We see Mysql's CPU temporarily saturates during the period of interest.

This situation highlights the need to create features to represent magnitudes like counters, percentages or rations and derivatives like velocity and acceleration. In the prior situation, we created features directly from data extracted from performance logs. In this case, Mysql flushing dirty pages to disk is a phase change. Diagnosing this type of millibottleneck requires features to account for a magnitude such as CPU utilization and a velocity measure like the change in dirty pages. As such, derivative measures are also components of the event and state feature matrices. In this illustration, the sudden change in the number of dirty pages is the primary signal. During this period of interest, we see this change corresponds to the other conditions present during this period: Mysql CPU suddenly and temporarily saturating, queue lengths elongating and the number of VLRT requests increasing.

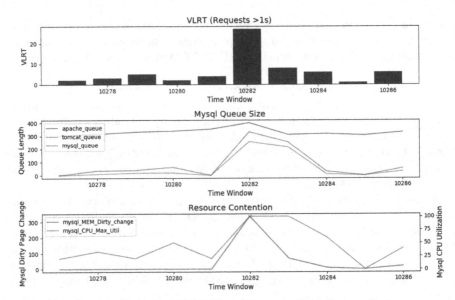

Fig. 4. Dirty Page as Millibottleneck Example (Color figure online)

This scenario also highlights the need to look for performance patterns across multiple resources across multiple components simultaneously. To account for this multiplicity, we employ a team-based classification approach to learning. Specifically, we train millibottleneck-specific models meaning we train over data containing negative and positive examples where the positive examples are of the same type. During prediction, we feed the feature vectors corresponding to a period of interest into each model. The model with the highest confidence is the final prediction for the given input. In this case, the model trained on data consisting of memory dirty page examples returns with the highest probability indicated by the red "X's" in the bottom figure.

4 megatables

Our system for automatically identifying millisecond-scale performance patterns at scale and with high confidence, *megatables*, consists of two primary components as shown in Fig. 5. First, our system automatically transforms event and resource monitoring log data into relational structures. Then, we transform the data "important" to detecting millibottlenecks into column feature vectors, which are assembled into an Event and State matrix. We use both matrices as inputs to a machine learning component composed of Teamed Classifiers. Our Team Classification design enables us to train system and pathology-specific models. Finally, megatables outputs millibottleneck detection predictions and supporting visualizations for any input period of interest.

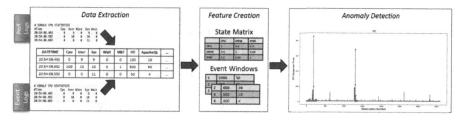

Fig. 5. Megatables System Overview. Request traces and performance (resource) logs are transformed into features, which are used by machine learning models to detect the presence of performance anomalies like millibottlenecks.

4.1 Millibottleneck Detection

megatables employs a Teamed Classifiers machine learning model to generate its millibottleneck predictions. Teamed Classifiers are an ensemble-based method. Models are arranged in layers, and each model is free to have its own form.

In our case, we employ random forests and decision trees as the form for each "team member," i.e. model instance. These forms seemed to best approximate our manual detection procedure, which is a combination of rules-based filtering and statistical correlation. Besides its lack of scalability, our previous method relied upon some subjective assessment. These non-parameteric forms provide a scalable mechanism for understanding, i.e. simple Boolean logic can explain the application of a label to a period.

We train a model to classify each type of millibottleneck. As highlighted in Sect. 2, we have identified approximately 10 different sources of millibottlenecks. Each model enables us to capture the millibottleneck-specific event and state dependencies across the systems' components. Decision trees and random forests, which are collections of decision trees, use tree-based data structures to partition data to minimize some information theoretic measure like entropy. These model forms are particularly robust for modeling non-linear relationships, and several of our millibottlenecks have non-linear components like phase transitions associated with buffer flushing. The numbers of trees and minimum number of samples per leaf are hyperparameters for these model forms. We achieved our best results using 100 trees and 2 samples per leaf for these hyperparameters.

Labeling Data. Given our treatment of millibottleneck detection as a supervised machine learning problem, our training data needs to contain labels. We label our data using two approaches. For certain classes of millibottlenecks, we can label the data using simple, deterministic rules like those induced by JVM 1.5 Garbage Collection. This prior enables us to label any temporary spike in CPU coinciding with regular garbage collection intervals as a millibottleneck. Our second approach amounts to manually labeling periods that we have previously diagnosed [7–10].

4.2 Feature Creation

megatables transforms raw performance data into salient numerical features for detecting millibottlenecks. We model systems with event and state matrices. These matrices are composed of event and state feature vectors. Our method for constructing these matrices must generalize and scale across different topologies, computer architectures, component (or application) software and resource monitors.

Event Matrix. We use request trace data to calculate metrics every 50 ms, specifically: Point-in-Time Response Time, Number of Very Long Response Time (VLRT) requests, and Queue Length for *each system component*. From these metrics, we create 100 ms windows and apply mean and max/min aggregate functions to windows values. Finally, we normalize each component of our vector using min-max scaling.

Besides accounting for different system topologies, our approach needs to account for different numbers of nodes in the topology, different numbers of CPUs on each node and different resources.

State Matrix. State features are obtained from CPU, Disk, Memory and Network resource monitoring data. As mentioned earlier, each resource on *each node* is represented as a multivariate time series. To accommodate this multiplicity, we employ fixed-sized windows and apply mean, max/min, and standard deviation to each variable. We also calculate first and second derivatives for each variable to account for each variable's change in velocity and acceleration over each period. Like our event matrix, we normalize each vector component using min-max scaling.

4.3 Data Extraction

Performance logs are semi-structured data but present some unique challenges relative to other semi-structured log data. Specifically, there are fewer semantics, i.e. limited metadata, and generally performance logs do not have fixed schemas. Consequently, layouts are not fixed *a priori* due to runtime factors such as system behavior, monitoring parameters, (i.e. which resources to monitor), and system architecture characteristics. Lastly, these logs can also have complex record structures such as containing multiple record types and degenerative sub-structures such as variable length record sizes. Prior work has assumed records are fixed size or occur within a fixed amount of space [11].

Figure 6 illustrates some of the challenges with extracting performance data automatically. The sections highlighted in Red correspond to *active processes* at the specified time. The sections highlighted in Blue correspond to *context switch* data. Each highlighted section represents different types of records, i.e. multiple record types A and B. Moreover, each record type has two records, i.e. A has 2 and B has 2. Lastly, A's records are variable length, i.e. 95 and 90 "rows" respectively. Note: the number in braces represents the number of intermediate rows, which have been removed for space.

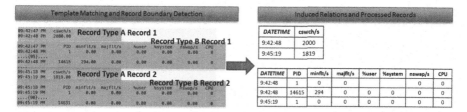

Fig. 6. Data Extraction and Relation Induction. Each of the shaded regions correspond to two different types of records. In this case, there are two instances of each record structure, which are extracted into two relations—one for each record type. (Color figure online)

To automatically extract data, megatables interprets files by matching log text to Layout Templates, specific but common layout-based patterns we have defined. Next, it induces a graph using information obtained from one (or more) Layout Templates' matching text. We developed a novel record boundary detection algorithm, which we describe later, irrespective of the record size. Finally, megatables uses the Layout Template and identified record boundaries to induce relations from the matching text.

5 Evaluation

We explore our method's diagnostic and extraction performance by assessing its reconstruction recall and coverage across two different datasets. The first data set comes from executing thousands of system benchmark experiments using a common, reddit-style bulletin board system. The second data set comes from randomly sampling thousands of performance log samples obtained from the source code website GitHub.

5.1 Reconstructing Distributions

We begin by assessing megatables ability to reconstruct the ground truth distributions for Point-in-Time Response Time and the number of VLRT request. As we have shown in the case examples earlier, these are strong indicators of the presence of millibottlenecks.

Figure 7 compares the ground truth Point-in-Time Response Time and VLRT distributions to their predicted equivalents. The figure provides strong visual evidence of our approach's ability to reconstruct these distributions across millibottleneck and non-millibottleneck periods. Secondly, we compare VLRT and Point-in-Time Response Time as indicators of the presence of millibottlenecks. Figure 8 compares the predicted data distributions for Point-in-Time response time and the number of VLRT requests split between millibottleneck and non-millibottleneck periods. These graphs suggest some interesting results. First, the number of VLRT requests provides better separability between the classes. For periods with no millibottlenecks, there is less than 0.01% weight in the tail suggesting that when there are no VLRT, millibottlenecks are not likely present. The Point-in-Time graphs suggest this metric does not separate the classes as well. Approximately 70% of the weight of the distribution occurs between 0 and 2000 for the Millibottleneck case. In the No Millibottleneck case, over 90% of the weight of the distribution occurs between these same thresholds. This overlap in the distributions suggests that a Point-in-Time prior can lead to millibottleneck misclassification.

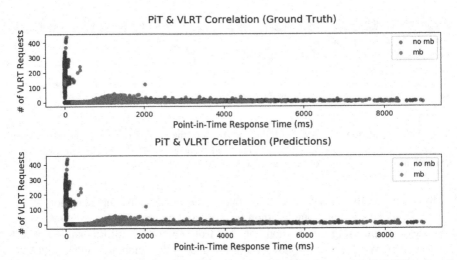

Fig. 7. VLRT and Point-in-Time comparison among predicted and ground truth Millibottleneck and No Millibottleneck periods

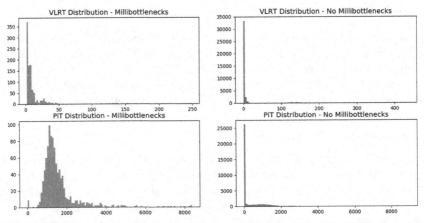

Fig. 8. VLRT and Point-in-Time Distribution comparisons for periods with Millibottlenecks and No Millibottlenecks detected

5.2 Data Extraction Coverage

We evaluate *megatables* coverage of performance monitor data by collecting a dataset from the wild via a popular public source code repository, GitHub. We eliminated those files from the sample that did not originate from performance monitoring programs. We retrieved this dataset by querying Github for keywords such as "log," "nagios," and "top." The latter two terms refer to two popular open source resource monitoring tools. Given their widespread use, we thought they should be included in our sample. Table 1 details our Github sample's characteristics.

We adopted Gao et al.'s record type categorization for describing log format/layout variety [11] with one modification. For files with interleaved record structures, we do not distinguish between those with single line and multiple line records. Instead, we use the number of interleaved record structures to explicitly illustrate their variety. Our sample covers at least 7 unique monitors not currently deployed in our infrastructure, including: top (profiling and processes), vmstat (vmware), oprofile, nagios, logstat (kvm) and a bespoke CPU/network monitoring tool.

Table 1. Github Sample Characteristics

Record Types	# of Samples	Average Size (# of Lines)
One (Single Line)	336	350
One (Multiple Lines)	434	120
Two or More	497	1559

Github Coverage. Figure 9 show *megatables* ability of our method to extend beyond the monitors used in our experimental computer science infrastructure. On average, we were able to correctly extract over 70% of the data obtained from GitHub into relations. Instances where most of the data could not be extracted were primarily due to our approach treating network message labels as attribute labels instead of elements of an enumeration. Despite this result, repairing this error can be accomplished with some simple post-processing.

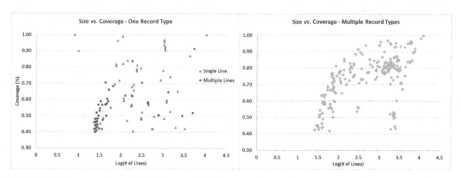

Fig. 9. Performance logs randomly sampled from Github. Each graph compares the coverage and size for files containing a Single Record Type (left) and Multiple Record Types (right).

Boundary Identification. We induce a graph, termed Record Boundary Graph (RBG), from the instances where log text matches a Layout Template. A vertex in the RBG corresponds to a template matching some text, termed an "instance." A directed edge is induced for vertices j → k if the instance corresponding to vertex j appears in the file (from top to bottom) before the instance corresponding to vertex k. After edges are induced, we find all simple cycles (or elementary circuits) in the graph. We bound the time complexity of this operation by limiting the number of visits a vertex can be visited to two. Next, we sort the identified simple cycles in descending order according to their length. We use the longest simple cycle identified in the RBG as the principle record boundary for defining candidate relations.

6 Related Work

Previous work on diagnosing performance bugs on distributed cloud systems generally follow a two-step approach. First, data is extracted from artifacts like source code or logs [12]. Then, these artifacts are interpreted with methods like machine learning or program analysis techniques to infer a system model. We describe previous work aligned to each of these steps; however, unlike megatables, few encompass both steps under one system.

Log Analysis. Several approaches employ machine learning to detect performance anomalies from log files [4, 12–15]. Generally, these approaches learn a model either online or offline from execution logs with performance bugs labeled accordingly. Xu et al. uses Principal Component Analysis (PCA) and critical thresholds to identify periods of anomalous system performance [12]. Du et al. uses a deep neural net and statistical outlier thresholds to detect performance anomalies [4]. One recent approach does not employ machine learning but instead relies on reconstructing programmers' event logging to profile system behavior [16].

Log Data Extraction. Approaches from previous work in automated information extraction has generally relied upon wrapper induction techniques. Specifically, this work has been applied to web data extraction using structural regularities among HTML tags to separate data from its presentation [17–19]. Work from the systems and programming language communities feature similar work on log data extraction. Some work has synthesized transformations to automatically generate a transformation program from user provided transformation actions. RecordBreaker is one such example of such an approach [20]. Other work from this community has relied on source code interposition techniques to decorate logging statements corresponding to specific strings inserted into the output [21]. Lastly, Datamaran uses parse trees to generate regular expressions, which are used to isolate log data structures [11].

7 Conclusion

Cloud computing has made it possible to decompose larger, on-premises monolithic applications into a set of smaller, atomistic distributed services called microservices that are deployable on public clouds such as AWS. This emerging cloud pattern represents the next step in enabling application owners to instantaneously scale their IT infrastructures according to their demand. While this evolution has enabled better economies of scale, it has come at the expense of performance predictability and ease of performance diagnosis. In this paper, we introduced our approach, megatables, for automatically detecting and analyzing fine-grained performance anomalies like millibottlenecks from disparate system logs. We demonstrated through three different scenarios that we can successfully use megatables to automatically detect millibottlenecks on different components due to different resource contention. Secondly, we showed its predictive ability and coverage by reconstructing the long tail latency distribution from a large catalog of systems' experiments. Finally, we demonstrated that our approach extends beyond the array of performance monitors present in our infrastructure by applying it to a large random sample of performance monitoring logs gathered from the wild.

References

1. Kohavi, R., Longbotham, R.: Online experiments: lessons learned. Computer **40**, 103–105 (2007)
2. Kohavi, R., Henne, R.M., Sommerfield, D.: Practical guide to controlled experiments on the web: listen to your customers not to the hippo. In: Proceedings of the 13th ACM SIGKDD International Conference on Knowledge Discovery and Data Mining (2007)
3. Pu, C., et al.: The millibottleneck theory of performance bugs, and its experimental verification. In: 2017 IEEE 37th International Conference on Distributed Computing Systems (ICDCS) (2017)
4. Du, M., Li, F., Zheng, G., Srikumar, V.: Deeplog: anomaly detection and diagnosis from system logs through deep learning. In: Proceedings of the 2017 ACM SIGSAC Conference on Computer and Communications Security (2017)
5. Gan, Y., et al.: Seer: leveraging big data to navigate the complexity of performance debugging in cloud microservices. In: Proceedings of the Twenty-Fourth International Conference on Architectural Support for Programming Languages and Operating Systems (2019)
6. Wang, Q., et al.: Lightning in the cloud: a study of transient bottlenecks on n-tier web application performance. In: 2014 Conference on Timely Results in Operating Systems ({TRIOS} 14) (2014)
7. Wang, Q., Kanemasa, Y., Kawaba, M., Pu, C.: When average is not average: large response time fluctuations in n-tier systems. In: Proceedings of the 9th International Conference on Autonomic Computing (2012)
8. Wang, Q., Kanemasa, Y., Li, J., Lai, C.A., Matsubara, M., Pu, C.: Impact of DVFS on n-tier application performance. In: Proceedings of the First ACM SIGOPS Conference on Timely Results in Operating Systems (2013)
9. Wang, Q., et al.: An experimental study of rapidly alternating bottlenecks in n-tier applications. In: 2013 IEEE Sixth International Conference on Cloud Computing (2013)
10. Park, J., Wang, Q., Li, J., Lai, C.-A., Zhu, T., Pu, C.: Performance interference of memory thrashing in virtualized cloud environments: a study of consolidated n-tier applications. In: 2016 IEEE 9th International Conference on Cloud Computing (CLOUD) (2016)
11. Gao, Y., Huang, S., Parameswaran, A.G.: Navigating the Data Lake with DATAMARAN - Automatically Extracting Structure from Log Datasets. SIGMOD Conference (2018)
12. Xu, W., Huang, L., Fox, A., Patterson, D., Jordan, M.I.: Detecting large-scale system problems by mining console logs. In: Proceedings of the ACM SIGOPS 22nd Symposium on Operating Systems Principles (2009)
13. Nagaraj, K., Killian, C., Neville, J.: Structured comparative analysis of systems logs to diagnose performance problems. In: Presented as part of the 9th USENIX Symposium on Networked Systems Design and Implementation (NSDI 12) (2012)

14. Stenzel, O.: and. The Physics of Thin Film Optical Spectra. SSSS, vol. 44, pp. 163–180. Springer, Cham (2016). https://doi.org/10.1007/978-3-319-21602-7_8
15. Zhang, K., Xu, J., Min, M. R., Jiang, G., Pelechrinis, K., Zhang, H.: Automated IT system failure prediction: a deep learning approach. In: 2016 IEEE International Conference on Big Data (Big Data) (2016)
16. Zhao, X., Rodrigues, K., Luo, Y., Yuan, D., Stumm, M.: Non-intrusive performance profiling for entire software stacks based on the flow reconstruction principle. In: 12th USENIX Symposium on Operating Systems Design and Implementation (OSDI 16) (2016)
17. Liu, L., Pu, C., Han, W.: XWRAP: an XML-enabled wrapper construction system for web information sources. In: Proceedings of 16th International Conference on Data Engineering (Cat. No. 00CB37073) (2000)
18. Han, W., Buttler, D., Pu, C.: Wrapping web data into XML. ACM SIGMOD Record **30**, 33–38 (2001)
19. Arasu, A., Garcia-Molina, H.: Extracting structured data from web pages. SIGMOD Conference (2003)
20. Fisher, K., Walker, D., Zhu, K.Q., White, P.: From dirt to shovels - fully automatic tool generation from ad hoc data. POPL (2008)
21. He, P., Zhu, J., Zheng, Z., Lyu, M.R.: Drain: an online log parsing approach with fixed depth tree. In: 2017 IEEE International Conference on Web Services (ICWS)

Keyphrase Extraction in Scholarly Digital Library Search Engines

Krutarth Patel[1]($\boxtimes$), Cornelia Caragea[2]($\boxtimes$), Jian Wu[3]($\boxtimes$), and C. Lee Giles[4]($\boxtimes$)

[1] Computer Science, Kansas State University, Manhattan, USA
kipatel@ksu.edu
[2] Computer Science, University of Illinois at Chicago, Chicago, USA
cornelia@uic.edu
[3] Computer Science, Old Dominion University, Norfolk, USA
jwu@cs.odu.edu
[4] Information Sciences and Technology, Pennsylvania State University,
State College, USA
giles@ist.psu.edu

Abstract. Scholarly digital libraries provide access to scientific publications and comprise useful resources for researchers who search for literature on specific subject areas. CiteSeerX is an example of such a digital library search engine that provides access to more than 10 million academic documents and has nearly one million users and three million hits per day. Artificial Intelligence (AI) technologies are used in many components of CiteSeerX including Web crawling, document ingestion, and metadata extraction. CiteSeerX also uses an unsupervised algorithm called noun phrase chunking (NP-Chunking) to extract keyphrases out of documents. However, often NP-Chunking extracts many unimportant noun phrases. In this paper, we investigate and contrast three supervised keyphrase extraction models to explore their deployment in CiteSeerX for extracting high quality keyphrases. To perform user evaluations on the keyphrases predicted by different models, we integrate a voting interface into CiteSeerX. We show the development and deployment of the keyphrase extraction models and the maintenance requirements.

Keywords: Scholarly digital libraries · Keyphrase extraction · Information extraction

1 Introduction

Online scholarly digital libraries usually contain millions of scientific documents [29]. For example, Google Scholar is estimated to have more than 160 million documents [38] Open access digital libraries have witnessed a rapid growth in their document collections as well in the past years [30]. For example, CiteSeerX's collection increased from 1.4 million to more than 10 million within the last decade. On one hand, these rapidly-growing scholarly document collections offer

© Springer Nature Switzerland AG 2020
W.-S. Ku et al. (Eds.): ICWS 2020, LNCS 12406, pp. 179–196, 2020.
https://doi.org/10.1007/978-3-030-59618-7_12

rich domain specific information for knowledge discovery, but, on the other hand, they pose many challenges to navigate and search for useful information in these collections.

Keyphrases of scientific papers provide important topical information about the papers in a highly concise form and are crucial for understanding the evolution of ideas in a scientific field [21,28,46]. In addition, keyphrases play a unique role in many downstream applications such as finding good index terms for papers [42], summarizing scientific papers [2,40,41], suggesting keywords in query formulation and expansion [45], recommending papers to readers [26], identifying reviewers for paper submissions [5], and clustering papers for fast retrieval [22]. Due to the high importance of keyphrases, several online digital libraries such as the ACM Digital Library have started to impose the requirement for author-supplied keyphrases. Specifically, these libraries require authors to provide keyphrases that best describe their papers. However, keyphrases have not been integrated into all sharing mechanisms. For example, the AAAI digital library (http://www.aaai.org/) does not provide keyphrases associated with the papers published in the AAAI conferences. In an effort to understand the coverage of papers with author-supplied keyphrases in open access scholarly digital libraries, we performed the following analysis: we randomly sampled 2,000 papers from CiteSeerX, and manually inspected each paper to determine whether a paper contains author-supplied keyphrases and if the paper was published by ACM. Note that in most of the ACM conference proceeding templates, the authors need to provide keyphrases (keywords) after the "Abstract" section. For completeness, the ACM templates from years 1998, 2010, 2015, and 2017 were adopted for visual inspection. Out of our 2,000 sample, only 31 (1.5%) papers were written using ACM templates and only 769 papers (38%) contain author-supplied keyphrases. Out of 31 papers written using ACM templates, 25 contain author-supplied keyphrases. The fact that around 62% of papers sampled do not have author-supplied keyphrases indicates that automatic keyphrase extraction is needed for scholarly digital libraries.

To date, many methods on the keyphrase extraction task have been proposed that perform better than NP-chunking or *tf-idf* ranking. Such methods include KEA [16], Hulth [27], TextRank [36], Maui [35], CiteTextRank [18], ExpandRank [49], CeKE [9], PositionRank [15], Key2Vec [34], BiLSTM-CRF [4], and CRFs based on word embeddings and document specific features [39]. However, keyphrase extraction has not been integrated into open access digital libraries. Most existing scholarly digital libraries [53] such as Google Scholar and Microsoft Academic do not display keyphrases. Recently, SemanticScholar started to display keyphrase-like terms called "topics." The CiteSeerX website currently displays keyphrases extracted using an unsupervised phrase chunking method [12].

In this application paper, we first review keyphrase extraction in scholarly digital libraries, using CiteSeerX as a case study. We investigate the impact of displaying keyphrases on promoting paper downloading by analyzing search engine access logs in three years from 2016 to 2018. Then, we interrogate the quality of several supervised keyphrase extraction models to explore their deployment

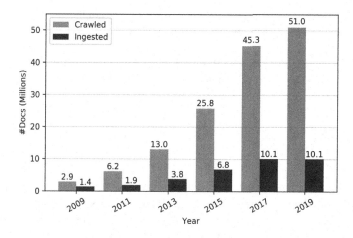

Fig. 1. Number of documents crawled and ingested from past few years in CiteSeerX.

in CiteSeerX and perform a large scale keyphrase extraction - first of its kind for this task. Moreover, to get user evaluations on the predicted keyphrases on a large scale, we implement and integrate a voting interface, which is widely used in social networks and multimedia websites, such as Facebook and YouTube. We show the development and deployment requirements of the keyphrase extraction models and the maintenance requirements.

2   CiteSeerX Overview and Motivation

There are in general two types of digital library search engines. The first type obtains publications and metadata from publishers, such as ACM Digital Library, IEEE Xplore, and Elsevier. The other type, such as CiteSeerX [17], crawls the public Web for scholarly documents and *automatically* extracts metadata from these documents.

CiteSeer was launched in 1998 [17] and its successor CiteSeerX [54] has been online since 2008. Since then, the document collection has been steadily growing (see Fig. 1). The goal of CiteSeerX is to improve the dissemination of and access to academic and scientific literature. Currently, CiteSeerX has 3 million unique users world-wide and is hit 3 million times a day. CiteSeerX reaches about 180 million downloads annually   [47]. Besides search capabilities, CiteSeerX also provides an Open Archives Initiative (OAI) protocol for metadata harvesting. CiteSeerX receives about 5,000 requests per month to access the OAI service. Researchers are interested in more than just CiteSeerX metadata. For example, CiteSeerX receives about 10 requests for data per month via the contact form on the front page [50]. These requests include graduate students seeking project datasets and researchers that were looking for large datasets for experiments. CiteSeerX hosts a dump of the database and other data on Google Drive.

In the early stage, the crawl seeds were mostly homepages of scholars in computer and information sciences and engineering (CISE). In the past decade,

Table 1. The number of full text documents, the total number of keyphrase-clicks, and unique keyphrases clicked for years 2016, 2017, and 2018 in CiteSeerX.

Year	#Docs. (Millions)	#Keyphrase-Clicks (Millions)	#Unique-Keyphrases (Millions)
2016	8.44	4.41	1.60
2017	10.1	7.17	1.86
2018	10.1	7.52	1.74

CiteSeerX added to the crawls seed URLs from the Microsoft Academic Graph [44], and directly incorporated PDFs from PubMed, arXiv, and digital repositories in a diverse spectrum of disciplines. A recent work on subject category classification of scientific papers estimated that the fractions of papers in physics, chemistry, biology, materials science, and computer science are 11.4%, 12.4%, 18.6%, 5.4%, and 7.6%, respectively [51]. CiteSeerX is increasing its document collection by actively crawling the Web using new policies and seeds to incorporate new domains. We expect this to encourage users from multiple disciplines to search and download academic papers and to be useful for studying cross discipline citation and social networks.

Since CiteSeerX was developed, many artificial intelligence techniques have been developed and deployed in CiteSeerX [54], including but not limited to header extraction [23], citation extraction [13], document type classification [11], author name disambiguation [48], and data cleansing [43]. In addition, an unsupervised NP-Chunking method is deployed for automatic keyphrase extraction. Besides author-submitted keyphrases, CiteSeerX extracts on average 16 keyphrases per paper using NP-Chunking. Users can search for a particular keyphrase by clicking it. This feature provides a shortcut for users to explore scholarly papers in related topics of the current paper they are browsing. All automatically extracted keyphrases are displayed on the summary page, and they deliver detailed domain knowledge in scholarly documents. Every time a keyphrase is clicked, CiteSeerX searches the clicked keyphrase and refreshes the search results. To understand how keyphrases promote paper browsing and downloading, we analyze the access logs retrieved from three web servers from 2016 to 2018.

2.1 Click-Log Analysis

Table 1 shows the total number of documents, keyphrase clicks, and unique keyphrases clicked from 2016 to 2018. The total number of keyphrase clicks increased significantly by ~63% from 2016 to 2017. For years 2017 and 2018, although the total number of documents stayed about the same (10.1 million), the total number of keyphrase clicks increased by 5%. Although there is a slight decrease in the number of unique keyphrases clicked, the increase in the number of keyphrase clicks from year 2016 to year 2018 showcases the increasing use and the popularity of keyphrases.

Figure 2 shows the ranking versus the number of clicks (#clicks) in logarithmic scale for the 10,000 most popular keyphrases during the three years. We can see that the #click decreases exponentially as the rank increases, which mimics the Zipf's law for all three years.

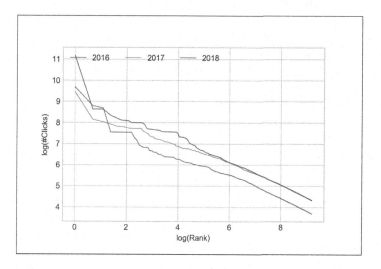

Fig. 2. log(Rank) vs log(Clicks) for top-10, 000 keyphrases clicked by users of CiteSeerX during years 2016, 2017, and 2018.

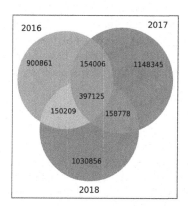

Fig. 3. Venn Diagram for all 3 years based on unique keyphrases.

Figure 3 shows the Venn diagram for the unique keyphrases clicked during •years 2016, 2017, and 2018. As seen from the figure, in two consecutive years,

Table 2. Top-20 keyphrases clicked during years 2016, 2017, and 2018.

Year	Keywords
2016	DgNe, local, bullying, violence, bullied, bully, aggressive, aggression, R. Nobrega, experimental result, data, wide range, machine, lpEu, dvd, last year, recent year, Artificial intelligence, key word, new technology
2017	Key word, experimental result, wide range, large number, string theory, bullying, Violence, bullied, bully, aggressive, aggression, recent year, new method, Artificial intelligence, important role, machine learning, neural network, Online version, environmental protection agency, wide variety
2018	JMQi, experimental result, key word, large number, wide range, aggression, Violence, bullying, bully, bullied, aggressive, recent year, case study, wide variety, Different type, sustainable development, informational security, VWBc, Sensor network, simulation result

only about one third of the keyphrases are common, whereas two third of the keyphrases are new. For example, 1.6 million unique keyphrases were clicked in 2016 but only about $551k$ (33%) were carried to 2017. Similarly, 1.86 million unique keyphrases were clicked in 2017, but only $555k$ (30%) were carried out in 2018. This trend implies that user interests have been rapidly evolving over these years, but there is still a considerable number of topics searched among several years. These conclusions are made based on the analysis of open-access documents from a three years time period. However, further analysis is needed for more comprehensive conclusions.

Table 2 shows the top-20 most frequent keyphrases clicked. We can see that the extracted keyphrases are not always terminological concepts as seen usually in author-submitted keyphrases. Examples such as "local", "experimental results", "wide range", and "recent year" were extracted just because they are noun phrases. This indicates that more sophisticated models are necessary to improve the quality of extracted keyphrases. It is interesting that these phrases were highly clicked, but investigating the reason is beyond the scope of this paper.

3 AI-Enabled Keyphrase Extraction

Here we describe three supervised keyphrase extraction models that we explore to integrate into CiteSeerX: KEA [16], Hulth [27], and Citation-enhanced Keyphrase Extraction (CeKE) [9]. Unlike KEA and Hulth, which only use the title and abstract of a given research article, CeKE exploits citation contexts along with the title and abstract of the given document. A citation context is defined as the text within a window of n words surrounding a citation mention. A citation context includes cited and citing contexts. A citing context for a target paper p is a context in which p is citing another paper. A cited context for a target paper p is a context in which p is cited by another paper. For a target paper, all cited contexts and citing contexts are aggregated into a single context.

Figure 4 shows an example of a small citation network using a paper (Paper 1) and its citation network neighbors. We can see the large overlap between the authors-submitted keyphrases and the citation contexts.

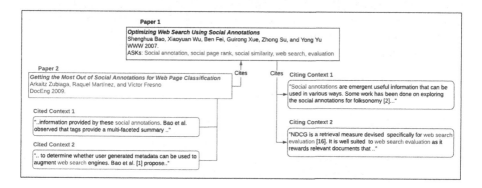

Fig. 4. A small citation network for Paper 1.

KEA: Frank et al. [16] used statistical features for the keyphrase extraction task and proposed a method named KEA. KEA uses following statistical features: *tf-idf*, i.e., the term frequency - inverse document frequency of a candidate phrase and the *relative position* of a candidate phrase, i.e., the position of the first occurrence of a phrase normalized by the number of words of the target paper. KEA extracts keyphrases from the title and abstract of a given paper.

Hulth: Hulth [27] argued that adding linguistic knowledge such as syntactic features can yield better results than relying only on statistics such as a term frequency (*tf*) and *n*-grams. Hulth showed remarkable improvement by adding part-of-speech (POS) tag as a feature along with statistical features. The features used in Hulth's approach are *tf*, *cf* (i.e., collection frequency), *relative position* and *POS tags* (if a phrase is composed by more than one word, then the POS will contain the tags of all words). Similar to KEA, Hulth extracts keyphrases only from the title and abstracts.

Citation-Enhanced Keyphrase Extraction (CeKE): Caragea et al. [9] proposed CeKE and showed that the information from the citation network in conjunction with traditional frequency-based and syntactical features improves the performance of the keyphrase extraction models.

CeKE uses the following features: *tf-idf* ; *relative position* ; POS tags of all the words in a phrase ; *first position* of a candidate phrase, i.e., the distance of the first occurrence of a phrase from the beginning of a paper; *tf-idf-Over*, i.e., a boolean feature, which is true if the *tf-idf* of a candidate phrase is greater than a threshold θ; *firstPosUnder*, also a boolean feature, which is true if the distance of the first occurrence of a phrase from the beginning of a target paper is below a certain threshold β. *Citation Network based features* include: *inCited* and *inCiting*, i.e., boolean features that are true if the candidate phrase occurs in cited and citing contexts, respectively; and *citation tf-idf*, i.e., the *tf-idf* score of each phrase computed from the aggregated citation contexts.

In our experiments, we compare three variants of CeKE: CeKE-Target that uses only the text from the target document; CeKE-Citing that uses the text from the target document and its citing contexts; CeKE-Cited that uses the text from the target document and its cited contexts; and CeKE-Both that uses both types of contexts.

Table 3. The dataset description.

ACM-CiteSeerX-KE					
Num. (#) Papers	Avg. # keyphrases	# keyphrases			
		#unigrams	#bigrams	#trigrams	# >trigrams
1,846	3.79	3,027	3,015	871	83

4   Experiments and Results

In this section, we first describe the dataset used for training and testing the keyphrase extraction models, the process of finding candidate phrases, and then present experimental results.

4.1   Dataset

We matched $30,000$ randomly selected ACM papers against all CiteSeerX papers by title and found $6,942$ matches. Among these papers, $6,942$, $5,743$, and $5,743$ papers have citing, cited, and both types of contexts, respectively. To create a dataset, we consider the documents for which we have both types of contexts and at least 3 author-supplied keyphrases appearing in titles or abstracts. We name this dataset as **ACM-CiteSeerX-KE**. Using these criteria, we identified 1,846 papers, which we used as our dataset for evaluation. The gold-standard contains the author-supplied keyphrases present in a paper (its title and abstract). Table 3 shows a summary of **ACM-CiteSeerX-KE** and contains the number of papers in the dataset, the average number of author-supplied keyphrases, and the number of n-gram author-supplied keyphrases, for $n = 1, 2, 3$, and $n > 3$.

4.2   Generating Candidate Phrases

We generate candidate phrases for each document by applying POS filters. Consistent with previous works [9,27,31,36,49], these candidate phrases are identified using POS-tags of words, consisting of only nouns and adjectives. We apply

Porter stemmer on each word. The initial position of each word is kept before removing any words. Second, to generate candidate phrases, contiguous words extracted in the first step are merged into n-grams ($n = 1, 2, 3$). Finally, we eliminate candidate phrases that end with an adjective and unigrams that are adjectives [9, 49].

Evaluation Metrics. To evaluate the performance of the keyphrase extraction methods, we use the following metrics: precision, recall and F1-score for the positive class since the correct identification of positive examples (keyphrases) is more important. These metrics are widely used in previous works [9, 27, 36, 49]. The reported values are averaged in 10-fold cross-validation experiments, where folds were created at document level and candidate phrases were extracted from the documents in each fold to form the training and test sets. In all experiments, we used Naïve Bayes on the feature vectors extracted by each model.

Table 4. The comparison of different models using 10-fold cross-validation on **ACM-CiteSeerX-KE**.

Model	Pr (%)	Re (%)	F1 (%)	Time/Doc (Sec)
NP-Chunking	04.26	29.19	07.44	**1.01**
Hulth	25.91	16.15	19.86	4.47
KEA	**30.41**	20.78	24.65	4.53
CeKE-Target	27.31	35.57	30.86	4.69
CeKE-Citing	25.65	40.45	31.37	6.61
CeKE-Cited	26.49	**42.73**	**32.68**	7.14
CeKE-Both	25.07	42.19	31.42	7.97

4.3 Results and Discussion

Table 4 shows the performance of NP-Chunking, KEA, Hulth, CeKE-Target, CeKE-Citing, CeKE-Cited, and CeKE-Both. The table shows the evaluation measures and time taken by each method using 10-fold cross-validation on **ACM-CiteSeerX-KE**. In NP-Chunking, the given text is first tokenized and tagged by a POS tagger. Based on the POS-tagging result, a grammar-based chunk parser is applied to separate two types of phrase chunks: (1) nouns or adjectives, followed by nouns (e.g., "relational database" or "support vector machine"), and (2) two chunks of (1) connected with a preposition or conjunction (e.g., "strong law of large numbers"). Time is measured on a computer with Xenon E5-2630 v4 processor and 32 GB RAM. In CiteSeerX, the header extraction tool can extract the title, abstract, and citing contexts for a target document. However, to extract cited contexts in CiteSeerX, there is an overhead of 1.2 s per document on average to search and extract it from the CiteSeerX database.

It can be seen from Table 4 that, CeKE-Cited achieves the highest recall and F1 of 42.73% and 32.68%, respectively. KEA achieves the highest precision of 30.41% compared with other models with top-5 predictions. NP-Chunking takes the shortest time of 1.01 s to extract keyphrases from a document. However, NP-Chunking suffers from low precision and F1. CeKE variants outperform Hulth and KEA in terms of recall and F1, i.e., CeKE-Citing achieves an F1 of 32.68% as compared with 24.65% achieved by KEA. Moreover, CeKE variants that make use of citation contexts outperform CeKE-Target that does not use any citation contexts.

It can be seen from the table that CeKE-Cited achieves highest F1 of 32.68%. However, CeKE-Citing takes less time compared with CeKE-Cited, i.e., CeKE-Citing takes 6.61 s on average per document compared with 7.14 s taken by CeKE-Cited. CeKE-Citing and CeKE-Both achieve comparable F1 of 31.37% and 31.42%, respectively. In terms of speed, CeKE-Target is the fastest among other variants because it does not need to perform POS tagging for citation contexts. Citing contexts can be extracted relatively straightforward from the content of the document. On the other hand, to extract cited contexts, we need the citation graph, from which we can obtain documents citing the target paper. We plan to select CeKE-Citing to deploy along with Hulth and KEA for the follow-

---

Title: Incorporating **site-level knowledge** to extract structured data from **web forums**

Abstract: Web forums have become an important data resource for many web applications, but extracting **structured data** from unstructured **web forum** pages is still a challenging task [...]. In this paper, we study the problem of **structured data** extraction from various **web forum** sites. Our target is to find a solution as general as possible to extract **structured data**, such as post title, post author, post time, and post content from any forum site. In contrast to most existing **information extraction** methods, which only leverage the knowledge inside an individual page, we incorporate both page-level and **site-level knowledge** and employ **Markov logic networks** (MLNs) [...]. The experimental results on 20 forums show a very encouraging **information extraction** performance, and demonstrate the ability of the proposed approach on various forums. [...]

Author-supplied keyphrases: *Web forums, Structured data, Information extraction, Site level knowledge, Markov logic networks*

CeKE-Citing predicted keyphrases: ***web forum, Site Level Knowledge**, forum, **structured data***
Hulth predicted keyphrases: *forum, page, Knowledge, post, **Site Level Knowledge**, **web forum, structured data***
KEA predicted keyphrases: ***Site Level Knowledge, web forum**, forum, post*

Fig. 5. The title, abstract, author-supplied keyphrases and predicted keyphrases of an ACM paper. The phrases marked with cyan in the title and abstract shown in the figure are author-supplied keyphrases.

ing reasons: CeKE-citing is faster than CeKE-cited and CeKE-Both; extracting cited contexts has an extra overhead to find it within a citation network; and cited context may not be present for all the articles.

Anecdotal Example: To demonstrate the quality of extracted phrases by different methods (CeKE-Citing, Hulth, and KEA), we select an ACM paper at random from the testing corpus and manually compared the keyphrases extracted by the three methods and the author-supplied keyphrases (Fig. 5). Specifically, the cyan bold phrases shown in the text on the top of the figure represent author-supplied keyphrases, whereas the bottom of the figure shows author-supplied keyphrases and predicted keyphrases by each evaluated model. It can be seen from the figure that the CeKE-Citing predicted four keyphrases out of which three are ASKs. Hulth predicted seven keyphrases out of which three are author-supplied keyphrases. KEA predicted three keyphrases out of which two belong to author-supplied keyphrases. The predicted keyphrases by all three models that do not belong to author-supplied keyphrases are single words. This example demonstrates that CeKE-citing exhibits a better performance than the other two models.

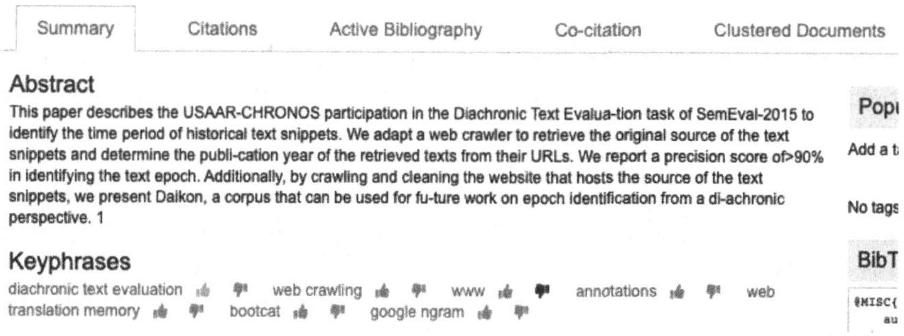

Fig. 6. A clip of a portion of a CiteSeerX paper's summary page containing a "Keyphrase" section that displays keyphrases extracted. Each keyphrase has a thumb-up and a thumbdown button. A logged in user can vote by clicking these buttons.

5   Crowd-Sourcing

The comparison between different keyphrase extraction models relies on ground truth datasets compiled from a small number of papers. We propose to evaluate keyphrase extraction models using crowd-sourcing, in which we allow users to vote for high quality keyphrases on papers' summary pages in CiteSeerX. These keyphrases are extracted using different models, but the model information is suppressed to reduce judgment bias. Voting systems are ubiquitous in social networks and multimedia websites, such as Facebook and YouTube, but they are rarely seen in scholarly digital libraries. A screenshot of an example of the voting interface is shown in Fig. 6. A database is already setup to store the total number of counts for each voting type as well as each voting action. The database contains the following tables.

- **Model table.** This table contains information of keyphrase extraction models.
- **Voting table.** This table contains the counts of upvotes and downvotes of keyphrases extracted using all models from all papers. The table also records the time the voting of a keyphrase is last updated. The same keyphrase extracted by two distinct models will have two entries in this table.
- **Action table.** This table contains information of all voting actions on keyphrases, such as the action time, the type of action (upvote vs. downvote), the IDs of keyphrases voted, and the IDs of voters. A voter must log in first before they can vote. If a voter votes a keyphrase extracted by two models, two actions will be recorded in this table. If a user reverses his vote, two actions (unvote and vote) are recorded in this table.

The extraction modules can be evaluated by the summation of eligible votes over all papers. In classic supervised machine learning, predicted keyphrases are evaluated by comparing extraction results against the author-supplied keyphrases [10]. However, the list of author-supplied keyphrases may not be exhaustive, i.e., certain pertinent keyphrases may be omitted by authors, but extracted by trained models. Crowd-sourcing provides an alternative approach that evaluates the pertinence of keyphrases from the readers' perspectives. However, there are certain potential biases that should be considered when deploying the system. One factor that can introduce bias is ordering because voters may not go through the whole list and vote all items. To mitigate this bias, we will shuffle keyphrases when displaying them on papers' summary pages. Another bias is the "Mathew's Effect" in which items with higher votes tend to receive more upvotes. We will hide the current votes of keyphrases to mitigate this effect.

We plan to collect votes after opening the voting system for at least 6 months. Using this approach, the keyphrase extraction models can be evaluated at two levels. At the *keyphrase level*, we only consider keyphrases with at least 10 votes and apply a binary judgment for keyphrase quality. A keyphrase is "favored" if the number of upvotes is higher than the downvotes, otherwise, it is labeled as "disfavored". We can then score each model based on the number of favored vs. disfavored. At the *vote level*, we can score each model using upvotes and downvotes of all keyphrases. The final scores should be normalized by the number of keyphrase extracted by a certain model and voted by users.

6   Development and Deployment

Although CiteSeerX utilizes open source software packages, many core components are not directly available from open source repositories and require extensive programming and testing. The current CiteSeerX codebase inherited little from its predecessor's (CiteSeer) for stability and consistency. The core part of the main web apps were written by Dr. Isaac Councill and Juan Pablo Fernández-Ramírez and many components were developed by other graduate students, postdocs and software engineers, which took at least 3–4 years.

CiteSeerX has been using keyphrases extracted using an unsupervised NP-Chunking method. This method is fast and achieves high recall, but it has a relatively low precision. Thus, we are exploring supervised models to extract keyphrases more accurately into CiteSeerX. Our keyphrase extraction module employs three methods: CeKE, Hulth, and KEA. The keyphrase extraction module runs on top of several dependencies, which handle metadata extraction from PDF files and document type classification in CiteSeerX. For example, GROBID [1] is used to extract titles, abstracts, and citing contexts. We also developed a program to extract cited contexts for a given article from the CiteSeerX database. In addition, a POS tagger[1] is a part of our keyphrase extraction module and is integrated in the keyphrase extraction module. Even though we selected CeKE-Citing, the keyphrase extraction package supports other variants of CeKE and it is straightforward to switch between them. Figure 7 shows the CiteSeerX system architecture and schematic diagram of our keyphrase extraction module.

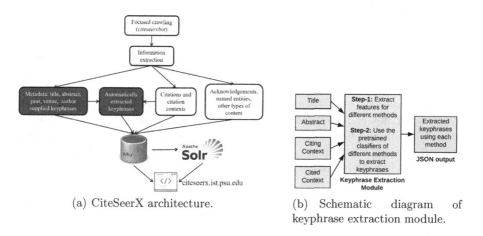

(a) CiteSeerX architecture.　　　　(b) Schematic   diagram   of keyphrase extraction module.

Fig. 7. CiteSeerX architecture and the keyphrase extraction module.

7   Maintenance

The keyphrase extraction module is developed and maintained by about 3 graduate students and a postdoctoral scholar in an academic setting. The keyphrase extraction project received partial financial support from the National Science Foundation. The maintenance work includes, but is not limited to fixing bugs, answering questions from GitHub users, updating extractors with improved algorithms, and rerunning new extractors on existing papers. Specific to the

---

[1] We have used NLP Stanford part of speech tagger.

keyphrase extraction module, it can easily integrate new models trained on different or large data for the existing methods. In future, we aim to integrate new keyphrase extraction models. The key bottleneck is to integrate keyphrase modules into the ingestion system, so both author-supplied keyphrases and predicted keyphrases can be extracted with other types of content at scale. One solution is to encapsulate keyphrase extraction modules into Java package files (.jar files) or Python libraries so they can easily be invoked by PDFMEF [52], a customizable multi-processing metadata extraction framework for scientific documents. Currently, the CiteSeerX group is developing a new version of digital library framework that employs PDFMEF as part of the information extraction pipeline. The encapsulation solution can potentially reduce the maintenance cost and increase modularity.

8 Related Work

Both supervised and unsupervised methods have been developed for keyphrase extraction [24]. These methods generally consists of two phases. In the first phase, candidate words or phrases are extracted from the text using heuristics such as POS patterns for words or n-grams [27]. In the second phase, the candidate phrases are predicted as keyphrases or non-keyphrases, using both supervised and unsupervised approaches.

In the supervised studies, keyphrase extraction is formulated as a binary classification problem or a sequential labeling. In the binary classification, the candidate phrases are classified as either keyphrase or non-keyphrase. In the sequential labeling, each token in a paper (sequence) is labeled as part of a keyphrase or not [4,19,39]. The prediction is done based on different features extracted from the text of a document, e.g., a word or phrase POS tags, *tf-idf* scores, and position information, used in conjunction with machine learning classifiers such as Naïve Bayes, Support Vector Machines, and Conditional Random Field [16,18,27,37]. The features extracted from external sources such as WordNet and Wikipedia [33,35]; from the neighbourhood documents, e.g., a document's citation network [8,9] were also used for the keyphrase extraction.

In unsupervised keyphrase extraction, the problem is usually formulated as a ranking problem. The phrases are scored using methods based on *tf-idf* and topic proportions [6,32,55]. The graph-based algorithms such as PageRank [20,36,49] and its variants [15,18,31] are also widely used in unsupervised models. Blank, Rokach, and Shani [7] ranked keyphrases for a target paper using keyphrases from the papers that are cited by the target paper and keyphrases from the papers that cite at least one paper that the target paper cites. The best performing model in SemEval 2010 [14] used term frequency thresholds to filter out unlikely phrases. Adar and Datta [3] extracted keyphrases by mining abbreviations from scientific literature and built a semantic hierarchical keyphrase database. Many of the above approaches, both supervised and unsupervised, are compared and analyzed in the ACL survey on keyphrase extraction by Hasan and Ng [25].

Usually, the performance of the supervised keyphrase extraction models is better than the unsupervised models [25].

9   Conclusions and Future Directions

By analyzing access logs of CiteSeerX in the past 3 years, we found that there are 3% of keyphrases common across all years, while there are many keyphrases which are only clicked during a particular year. In this application paper, we proposed to integrate three supervised keyphrase extraction models into CiteSeerX which are more robust than the previously used NP-Chunking method. To evaluate the keyphrase extraction methods from a user perspective, we implemented a voting system on papers' summary pages in CiteSeerX to vote on predicted phrases without showing the model information to reduce potential judgment bias from voters.

In the future, it would be interesting to integrate other keyphrase extraction models as well as other information extraction tools such as name-entity extraction tool to improve the user experience.

Acknowledgements. We thank the National Science Foundation (NSF) for support from grants CNS-1853919, IIS-1914575, and IIS-1813571, which supported this research. Any opinions, findings, and conclusions expressed here are those of the authors and do not necessarily reflect the views of NSF. We also thank our anonymous reviewers for their constructive feedback.

References

1. Grobid. https://github.com/kermitt2/grobid (2008–2020)
2. Abu-Jbara, A., Radev, D.: Coherent citation-based summarization of scientific papers. In: ACL: HLT, pp. 500–509 (2011)
3. Adar, E., Datta, S.: Building a scientific concept hierarchy database (schbase). In: ACL, pp. 606–615 (2015)
4. Alzaidy, R., Caragea, C., Giles, C.L.: Bi-lstm-crf sequence labeling for keyphrase extraction from scholarly documents. In: WWW, pp. 2551–2557. ACM (2019)
5. Augenstein, I., Das, M., Riedel, S., Vikraman, L., McCallum, A.: Semeval 2017 task 10: Scienceie-extracting keyphrases and relations from scientific publications. arXiv preprint arXiv:1704.02853 (2017)
6. Barker, K., Cornacchia, N.: Using noun phrase heads to extract document keyphrases. In: Hamilton, H.J. (ed.) AI 2000. LNCS (LNAI), vol. 1822, pp. 40–52. Springer, Heidelberg (2000). https://doi.org/10.1007/3-540-45486-1_4
7. Blank, I., Rokach, L., Shani, G.: Leveraging the citation graph to recommend keywords. In: RecSys, pp. 359–362 (2013)
8. Bulgarov, F., Caragea, C.: A comparison of supervised keyphrase extraction models. In: WWW, pp. 13–14 (2015)
9. Caragea, C., Bulgarov, F., Godea, A., Gollapalli, S.D.: Citation-enhanced keyphrase extraction from research papers: a supervised approach. In: EMNLP (2014)
10. Caragea, C., Bulgarov, F.A., Godea, A., Gollapalli, S.D.: Citation-enhanced keyphrase extraction from research papers: a supervised approach. In: Proceedings of the 2014 Conference on Empirical Methods in Natural Language Processing, EMNLP 2014, 25–29 October 2014, Doha, Qatar, A meeting of SIGDAT, a Special Interest Group of the ACL, pp. 1435–1446 (2014) http://aclweb.org/anthology/D/D14/D14-1150.pdf

11. Caragea, C., Wu, J., Gollapalli, S.D., Giles, C.L.: Document type classification in online digital libraries. In: Twenty-Eighth IAAI Conference (2016)
12. Chen, H.H., Treeratpituk, P., Mitra, P., Giles, C.L.: Csseer: an expert recommendation system based on citeseerx. In: JCDL, pp. 381–382 (2013)
13. Councill, I., Giles, C.L., Kan, M.Y.: ParsCit: an open-source CRF reference string parsing package. LREC **8**, 661–667 (2008)
14. El-Beltagy, S.R., Rafea, A.: Kp-miner: participation in semeval-2. In: SemEval, pp. 190–193 (2010)
15. Florescu, C., Caragea, C.: Positionrank: an unsupervised approach to keyphrase extraction from scholarly documents. In: ACL, pp. 1105–1115 (2017)
16. Frank, E., Paynter, G.W., Witten, I.H., Gutwin, C., Nevill-Manning, C.G.: Domain-specific keyphrase extraction. In: IJCAI, pp. 668–673 (1999)
17. Giles, C.L., Bollacker, K.D., Lawrence, S.: Citeseer: an automatic citation indexing system. In: JCDL, pp. 89–98 (1998)
18. Gollapalli, S.D., Caragea, C.: Extracting keyphrases from research papers using citation networks. In: AAAI, pp. 1629–1635 (2014)
19. Gollapalli, S.D., Li, X.L., Yang, P.: Incorporating expert knowledge into keyphrase extraction. In: AAAI, pp. 3180–3187 (2017)
20. Grineva, M., Grinev, M., Lizorkin, D.: Extracting key terms from noisy and multitheme documents. In: WWW, pp. 661–670 (2009)
21. Hall, D., Jurafsky, D., Manning, C.D.: Studying the history of ideas using topic models. In: EMNLP, pp. 363–371 (2008)
22. Hammouda, K.M., Matute, D.N., Kamel, M.S.: CorePhrase: keyphrase extraction for document clustering. In: Perner, P., Imiya, A. (eds.) MLDM 2005. LNCS (LNAI), vol. 3587, pp. 265–274. Springer, Heidelberg (2005). https://doi.org/10.1007/11510888_26
23. Han, H., Giles, C.L., Manavoglu, E., Zha, H., Zhang, Z., Fox, E.A.: Automatic document metadata extraction using support vector machines. In: JCDL, pp. 37–48. IEEE (2003)
24. Hasan, K.S., Ng, V.: Conundrums in unsupervised keyphrase extraction: making sense of the state-of-the-art. In: COLING, pp. 365–373 (2010)
25. Hasan, K.S., Ng, V.: Automatic keyphrase extraction: a survey of the state of the art. In: ACL, pp. 1262–1273, June 2014
26. Hong, K., Jeon, H., Jeon, C.: Personalized research paper recommendation system using keyword extraction based on userprofile. In: Journal of Convergence Information Technology (JCIT) (2013)
27. Hulth, A.: Improved automatic keyword extraction given more linguistic knowledge. In: EMNLP (2003)
28. Jurgens, D., Kumar, S., Hoover, R., McFarland, D., Jurafsky, D.: Measuring the evolution of a scientific field through citation frames. TACL **6**, 391–406 (2018)
29. Khabsa, M., Giles, C.L.: The number of scholarly documents on the public web. PLoS One **9**(5), 25 (2014)
30. Larsen, P., Von Ins, M.: The rate of growth in scientific publication and the decline in coverage provided by science citation index. Scientometrics **84**(3), 575–603 (2010)
31. Liu, Z., Huang, W., Zheng, Y., Sun, M.: Automatic keyphrase extraction via topic decomposition. In: EMNLP, pp. 366–376 (2010)
32. Liu, Z., Li, P., Zheng, Y., Sun, M.: Clustering to find exemplar terms for keyphrase extraction. In: EMNLP, pp. 257–266 (2009)
33. Lopez, P., Romary, L.: Humb: automatic key term extraction from scientific articles in grobid. In: SemEval, pp. 248–251 (2010)

34. Mahata, D., Kuriakose, J., Shah, R.R., Zimmermann, R.: Key2vec: automatic ranked keyphrase extraction from scientific articles using phrase embeddings. In: NAACL, pp. 634–639 (2018)
35. Medelyan, O., Frank, E., Witten, I.H.: Human-competitive tagging using automatic keyphrase extraction. In: EMNLP, pp. 1318–1327 (2009)
36. Mihalcea, R., Tarau, P.: Textrank: bringing order into texts. In: EMNLP (2004)
37. Nguyen, T.D., Kan, M.-Y.: Keyphrase extraction in scientific publications. In: Goh, D.H.-L., Cao, T.H., Sølvberg, I.T., Rasmussen, E. (eds.) ICADL 2007. LNCS, vol. 4822, pp. 317–326. Springer, Heidelberg (2007). https://doi.org/10.1007/978-3-540-77094-7_41
38. Orduña-Malea, E., Ayllón, J.M., Martín-Martín, A., López-Cózar, E.D.: Methods for estimating the size of google scholar. Scientometrics **104**(3), 931–949 (2015)
39. Patel, K., Caragea, C.: Exploring word embeddings in CRF-based keyphrase extraction from research papers. In: K-CAP, pp. 37–44. ACM (2019)
40. Qazvinian, V., Radev, D.R.: Scientific paper summarization using citation summary networks. In: COLING. pp. 689–696, Manchester, United Kingdom (2008)
41. Qazvinian, V., Radev, D.R., Özgür, A.: Citation summarization through keyphrase extraction. In: COLING, pp. 895–903 (2010)
42. Ritchie, A., Teufel, S., Robertson, S.: How to find better index terms through citations. In: CLIIR, pp. 25–32 (2006)
43. Sefid, A., et al.: Cleaning noisy and heterogeneous metadata for record linking across scholarly big datasets. In: The Thirty-Third AAAI Conference on Artificial Intelligence, AAAI 2019, The Thirty-First Innovative Applications of Artificial Intelligence Conference, IAAI 2019, The Ninth AAAI Symposium on Educational Advances in Artificial Intelligence, EAAI 2019, Honolulu, Hawaii, USA, January 27 - February 1, 2019, pp. 9601–9606 (2019)
44. Sinha, A., et al.: An overview of microsoft academic service (mas) and applications. In: WWW, pp. 243–246 (2015)
45. Song, I.Y., Allen, R.B., Obradovic, Z., Song, M.: Keyphrase extraction-based query expansion in digital libraries. In: JCDL, pp. 202–209 (2006)
46. Tan, C., Card, D., Smith, N.A.: Friendships, rivalries, and trysts: Characterizing relations between ideas in texts. arXiv preprint arXiv:1704.07828 (2017)
47. Teregowda, P., Urgaonkar, B., Giles, C.L.: Cloud 2010. In: 2010 IEEE 3rd International Conference on Cloud Computing, pp. 115–122 (2010)
48. Treeratpituk, P., Giles, C.L.: Disambiguating authors in academic publications using random forests. In: JCDL, pp. 39–48. ACM (2009)
49. Wan, X., Xiao, J.: Single document keyphrase extraction using neighborhood knowledge. AAAI. **8**, 855–860 (2008)
50. Williams, K., Wu, J., Choudhury, S.R., Khabsa, M., Giles, C.L.: Scholarly big data information extraction and integration in the citeseer digital library. IIWeb, pp. 68–73 (2014)
51. Wu, J., Kandimalla, B., Rohatgi, S., Sefid, A., Mao, J., Giles, C.L.: Citeseerx-2018: a cleansed multidisciplinary scholarly big dataset. In: IEEE Big Data, pp. 5465–5467 (2018)
52. Wu, J., et al.: Pdfmef: a multi-entity knowledge extraction framework for scholarly documents and semantic search. In: K-CAP, pp. 13:1–13:8. ACM (2015)

53. Wu, J., Liang, C., Yang, H., Giles, C.L.: Citeseerx data: Semanticizing scholarly papers. In: SBD, pp. 2:1–2:6. ACM (2016)
54. Wu, J., et al.: CiteSeerX: AI in a digital library search engine. In: AAAI, pp. 2930–2937 (2014)
55. Zhang, Y., Milios, E., Zincir-Heywood, N.: A comparative study on key phrase extraction methods in automatic web site summarization. JDIM **5**(5), 323 (2007)

Scheduling Multi-workflows over Edge Computing Resources with Time-Varying Performance, A Novel Probability-Mass Function and DQN-Based Approach

Hang Liu[1], Yuyin Ma[1], Peng Chen[2(✉)], Yunni Xia[1(✉)], Yong Ma[3], Wanbo Zheng[4], and Xiaobo Li[5]

[1] Software Theory and Technology Chongqing Key Lab, Chongqing University, Chongqing 400044, China
xiayunni@hotmail.com
[2] School of Computer and Software Engineering, Xihua University, Chengdu 610039, China
chenpeng@gkhb.com
[3] College of Computer Information Engineering, Jiangxi Normal University, Nanchang 330022, China
[4] Kunming University of Science and Technology, Kunming 650000, China
[5] Chongqing Animal Husbandry Techniques Extension Center, Chongqing 400044, China

Abstract. The edge computing paradigm is featured by the ability to off-load computing tasks from mobile devices to edge clouds and provide high cost-efficient computing resources, storage and network services closer to the edge. A key question for workflow scheduling in the edge computing environment is how to guarantee user-perceived quality of services when the supporting edge services and resources are with unstable, time-variant, and fluctuant performance. In this work, we study the workflow scheduling problem in the multi-user edge computing environment and propose a Deep-Q-Network (DQN) -based multi-workflow scheduling approach which is capable of handling time-varying performance of edge services. To validate our proposed approach, we conduct a simulative case study and compare ours with other existing methods. Results clearly demonstrate that our proposed method beats its peers in terms of convergence speed and workflow completion time.

Keywords: Workflow scheduling · Edge computing · Probability distribution model · Reinforcement learning · Deep Q network

1 Introduction

The edge computing paradigm is emerging as a high performance computing environment with a large-scale, heterogeneous collection of autonomous systems

H. Liu and Y. Ma—Contribute equally to this article and should be considered co-first authors.

© Springer Nature Switzerland AG 2020
W.-S. Ku et al. (Eds.): ICWS 2020, LNCS 12406, pp. 197–209, 2020.
https://doi.org/10.1007/978-3-030-59618-7_13

and flexible computational architecture [1–6]. It provides the tools and technologies to build data or computational intensive parallel applications with much more affordable prices compared to traditional parallel computing techniques. Hence, there has been an increasingly growth in the number of active research work in edge computing such as scheduling, placement, energy management, privacy and policy, security, etc. Workflow scheduling in cloud and edge environment has recently drawn enormous attention thanks to its wide application in both scientific and economic areas. A workflow is usually formulized as a Directed-Acyclic-Graph (DAG) with several n tasks that satisfy the precedent constraints. Scheduling workflows over an edge environment is referred to as matching tasks onto edge services created on edge nodes. For multi-objective scheduling, objectives can sometimes be conflicting. e.g., for execution time minimization, fast services are more preferable than slow ones. However, fast services are usually more expensive and thus execution time minimization may contradict the cost reduction objective. It is widely acknowledged as well that to scheduling multi-task workflow on distributed platforms is an NP-hard problem. It is therefore extremely time-consuming to yield optimal schedules through traversal-based algorithms. Fortunately, heuristic and meta-heuristic algorithms with polynomial complexity are able to produce approximate or near optimal solutions of schedules at the cost of acceptable optimality loss [7–12]. Good examples of such algorithms are multi-objective particle swarm optimization (MOPSO) and non-dominated sorting genetic algorithm-II (NSGA-II).

Recently, as novel machine learning algorithms are becoming increasingly versatile and powerful, considerable research efforts are paid to using reinforcement learning (RL) and Q-learning-based algorithms [13–15] in finding near-optimal workflow scheduling solutions. Nevertheless, most existing contributions in this direction focused on scheduling workflows over centralized clouds. How to apply Q-learning-based algorithms and models to the problem of scheduling workflows upon distributed edge computing platforms is still to be clearly addressed. In this work, we propose a DQN-based multi-workflow scheduling method. The proposed model takes the probability mass functions (PMF) of historical performance data of edge services as the inputs and is capable of improving the scheduling plans via optimizing the probability of a workflow satisfying the completion-time constraint. We conduct a simulated experiment and compare our method with other baseline algorithms. The results show that our method outperforms baseline algorithms in terms of workflow completion time.

2   Related Work

Scheduling multi-workflows upon distributed infrastructures, e.g., grids, clouds and edge, is usually known to be NP-hard and thus traversal-based algorithms can be ineffective in terms of computational complexity. Instead, heuristic and meta-heuristic procedures with polynomial complexity can yield high-quality and sub-optimal solutions at the cost of a certain level of optimality degradation. For example, [16] leveraged a multi-objective bio-inspired procedure (MOBFOA) by augmenting the tradtional BFOA with Pareto-optimal fronts. Their method deals with the reduction of flow-time, completion duration, and operational cost. [17] considered a multi-objective genetic optimization (BOGA) and optimizd both electricity consumption and DAG reliability. [18] considered an augmented GA with the Efficient Tune-In (GA-ETI) mechanism for the optimization of turnaround time. [19] employed a non-dominated-sorting-based Hybrid PSO approach and aimed at minimizing both turnaround time and cost. [20] introduced a fuzzy dominance sort based heterogeneous finishing time minimization approach for the optimization of both cost and turnaround time of DAG executed on IaaS clouds.

Recently, deep reinforcement learning (DRL) methods shed new light on the problem we are interested in [21–27]. It was shown that the multi-agent training methods can be effective in dealing with multi-constraint and multi-objective optimization problems. For example, [28] employed a sequential cooperative game approach for heterogeneous workflow scheduling. [29] developed a reinforcement-learning-based method for multi-DAG execution with user-defined priorities specified at different times. [30] proposed a distributed load management and access control approach for the SaaS environment by using a fuzzy game-theoretic model. [31] proposed modified Q-learning method for turn-around time reduction and load balancing by considering weighted fitness value function. However, Q-learning-based algorithms and models intended for edge-infrastructure-based workflow scheduling is very rare. A highly efficient reinforcement-learning-based approach for scheduling and managing multi-workflows upon distributed, mobile, and resource-constrained edge services is in high need.

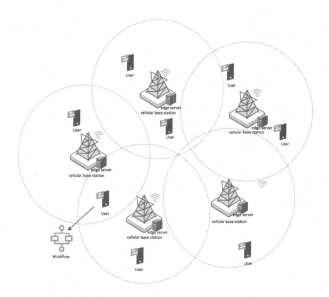

Fig. 1. Edge computing environment

3   Model and System

3.1   System Architecture

As shown in Fig. 1, an edge computing environment can be seen as a collection of multiple edge servers usually deployed near base stations. By this way, users are allowed to off-load compute-intensive and latency-sensitive applications, e.g., Augmented Reality (AR), Virtual Reality (VR), Artificial Intelligence (AI), to edge servers. Within an edge computing environment, there exist n users in an edge computing environment, denote by $U = \{u_1, u_2, ..., u_n\}$, and m base stations, denote by $B = \{b_1, b_2, ..., b_m\}$. Each user has an application to be executed, and users mobile device is allowed to offload tasks on edge servers near the base station by wireless access point. For generality, we regard mobile applications as workflows, denote by a directed acyclic graph(DAG) $W = (T, D)$, where $T = \{t_1, t_2, ..., t_n\}$ represents a set of tasks. Tasks have multiple types which have different size of input data. $D = \{d_{i,j}|i, j \in [1, n]\}$ represents a set of precedence dependencies, where $d_{i,j} = 1$ means t_j can be executed only when t_i is completed, otherwise $d_{i,j} = 0$. $S_i = \{s_1, s_2, ..., s_n\}$ represents the list of servers which signal coverage covers user i, thus user i can offload tasks on these servers.

Users are allowed to offload tasks to the edge via wireless access points. The action profile of users can be expressed as $a_i = \{s_1, s_2, ..., s_m\}$, where s_j indicates server s_j. For a server s_j, a list of users who offload tasks to it can be represented as $UL_j = \{i|s_j \in a_i\}$. For an action profile $A = \{a_1, a_2, ..., a_n\}$ of all users, the uplink data rate of wireless channel of user u_i to server s_j can be estimated by

$$R_{i,j}(A) = B \cdot log_2(1 + \frac{p_i g_{i,j}}{\sum_{k \in UL_j} p_k g_{k,j} + \sigma}) \tag{1}$$

where B is the channel bandwidth, p_i the transmit power of user u_i, $g_{i,j}$ the channel gain from user u_i to server s_j, and σ the backgroud noise power. It can thus be seen from this equation, if too many users choose to offload its tasks to the same server, the uplink data rate decreases and further causes low offloading efficiency.

Assume user u_i chooses to offload its task t_j to server s_k, according to the Eq. 1, the transmission time for offloading the input data of size $C_{i,j,k}$ can be estimated by

$$TT_{i,j,k}(A) = \frac{C_{i,j,k}}{R_{i,j}} = \frac{C_{i,j,k}}{Blog_2(1 + \frac{p_i g_{i,j}}{\sum_{k \in UL_j} p_k g_{k,j} + \sigma})} \tag{2}$$

We assume that all wireless channels obey the quasi-static block fading rule [32]. This rule means that the state of the channel remains unchanged during transmission. Thus, the probability distribution of the completion time of the task is

$$T_{i,j,k} = TT_{i,j,k}(A) + TE_{i,j,k} \tag{3}$$

$$PMF_{i,j,k}^{TE}(t) = Prob(TE_{i,j,k}) \tag{4}$$

$$Prob(T_{i,j,k}) = PMF^T_{i,j,k}(t) = PMF^{TE}_{i,j,k}(t - TT_{i,j,k}) \tag{5}$$

where TE is historical execution time, $PMF(t)$ indicates the probability mass function of the historical execution time.

3.2 Promblem Formulation

Based on the above system model, we are interested in knowing the highest probability of meeting the completion-time constraints. The resulting scheduling problem can be described as follows:

$$\max f = Prob_{avg} = \frac{1}{N} \sum_{i=1}^{N} Pr(CT_i <= C_i^g) \tag{6}$$

subject to,

$$i \in [1, N], CT_i \geq 0, C_i^g \geq 0 \tag{7}$$

where C_i^g is a completion-time threshold for user u_i and CT_i the actual completion time of a user's workflow.

4 Our Approach

4.1 Decomposition of the Global Constraint

For the evaluation of effectiveness of the actions by agents during the training process, we first have to decompose the global constraint to local ones. Given a workflow with n tasks, denoted by $W = \{t_1, t_2, ..., t_n\}$ and C^g as the global completion-time constraint, the local constraint of subtask can be represented by $C^l = \{C_1^l, C_2^l, ..., C_n^l\}$. We consider dividing the global constraint in proportion to the expected completion time of each part, where specific steps are as follows:

1. Obtain the server list whose coverage reach user k, denoted by $S_k = \{s_1, s_2, ..., s_n\}$.
2. For task t_i, its completion time on server s_j is represented by a PMF that we mentioned above. The expected completion time $e_{i,j}$ can be estimated by $\{e_{i,j}| \int_0^{e_{i,j}} PMF(X) = 0.5\}$.
3. Task t_i has multiple candidate servers S_k to be scheduled into, the expected completion time of task t_i is $E_i^t = avg(e_i)$, where $e_i = \{e_{i,1}, e_{i,2}, ..., e_{i,n}\}$.
4. For any part p_g, it consists of tasks $T_g = \{t_1, t_2, ..., t_n\}$. The expected completion time of this part is thus $E_g^p = \max(E_{t_1}^t, E_{t_2}^t, ..., E_{t_n}^t)$
5. Eventually, we can divide the global constraint into smaller ones as follows:

$$C_i^l = C^g \cdot \frac{E_i^p}{\sum_{j=1}^n E_j^p} \tag{8}$$

4.2   Deep-Q-Network-based Solution to the Workflow Scheduling Problem

As mentioned earlier, we employ DQN for solving the optimization formulations given above. According to DQN, the value function updated by time difference can be expressed as:

$$Q(s,a) = (1 - \alpha)Q(s,a) + \alpha[R(a) + \gamma \max_{a' \in A} Q(s',a')] \tag{9}$$

where $Q(s,a)$ is the state-action value function at current state, $Q(s',a')$ is the state-action value function at the next state, α is the update step size, $R(a)$ is the reward derived based on the PMF of the workflow completion time according to (12) and γ is the reward decay factor. The loss function of deep q network can be computed by

$$L(\theta) = E_{s,a,r,s'}[(Q^*(s,a|\theta) - y)^2] \tag{10}$$

$$y = R(a) + \delta \max_{a' \in A} Q^*(s',a') \tag{11}$$

where y presents the target Q network whose parameters are periodically replaced by evaluate Q network Q^*. The DQN procedure is shown in Algorithm 1.

Algorithm 1. Deep Q Learning algorithm

Initialize replay memory D to capacity N
Initialize action-value function Q with random weights
for *episode* $= 1, M$ **do**
    Initialise sequence $s_1 = \{x_1\}$ and preprocessed sequenced $\phi = \phi(s_i)$
    for $t = 1, T$ **do**
        With probability *epsilon* select a random action a_t
        otherwise select $a_t = \max_a Q^*(\phi(s_t), a; \theta)$
        Execute action a_t in emulator and observe reward r_t and state s_t
        Store transition $(\phi_t, a_t, r_t, \phi_{t+1})$ in D
        Sample random minibatch of transitions $(\phi_t, a_t, r_t, \phi_{t+1})$ from D
        Set $y_j = \begin{cases} r_j, & terminal\phi_{j+1} \\ r_j + \gamma \max_{a'} Q(\phi_{j+1}, a'; \theta), & non - terminal\phi_{j+1} \end{cases}$
        Perform a gradient descent step on $(y_j - Q(\phi_j, a_j; \theta)^2)$
    end for
end for

The DQN environment includes components of environment observation, action space, policy setting, and reward design [33]. Note that the former 3 components can be implemented by using the standard DQN setting, while the reward design one should be developed based on the optimization formulation and the constraint decomposition configuration given in the previous sections. The reward function is designed as:

$$R_i(a) = Pr(X \leq C_i^l)^3 \tag{12}$$

where C_i^l is based on the decomposition of the global constraint given in (8).

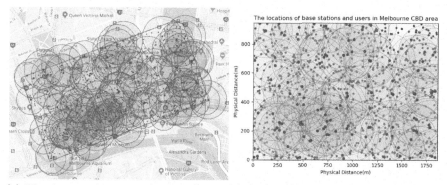

(a) The dataset shown on the Google map (b) The dataset shown on the coordinates

Fig. 2. An example of edge servers with their coverages and edge users in Melbourne BCD

5   Case Study

In this section, we conduct simulative case studies to prove the effectiveness, in terms of workflow completion time, network loss value, and convergence speed of the algorithm of our method. The types of server, workflow and task are randomly generated. We assume as well that edge servers and users are located according to the position dataset of [34] as illustrated in Fig. 2. Edge servers have 3 different types, i.e., *type*1, *type*2, and *type*3, in terms of their resource configuration and performance. User applications are expressed in the form of multiple workflows as given in Fig. 3, where each workflow task is responsible for executing a GaussCLegendre calculation with 8, 16, or 32 million decimal digits. The historical execution time for GaussCLegendre calculations over different types of edge servers are based on data from [35] shown in Fig. 4. For the comparison purpose, we compare our proposed method with other existing methods, i.e., NSPSO [36] and NSGA [37] as well.

5.1   Experiment Configuration

We test our methods and its peers by using a workstation with the Intel Core i7 CPU @ 2.80 GHz, NVIDIA GeForce GTX 1050 Ti, and 8 GB RAM configuration. Table 1 shows basic parameters used in the experiments.

5.2   Performance Evaluation

Based on the above configurations and datasets, we repeated invoking our proposed method to schedule workflows based on performance data of edge servers measured at 3 different time periods given in Fig. 4. It can be seen from Figs. 5 and 6 that the network loss decreases rapidly with time and the probability of satisfying global constraint increases with iterations.

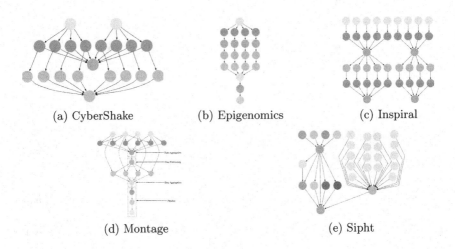

(a) CyberShake   (b) Epigenomics   (c) Inspiral

(d) Montage   (e) Sipht

Fig. 3. Five typical workflow templates

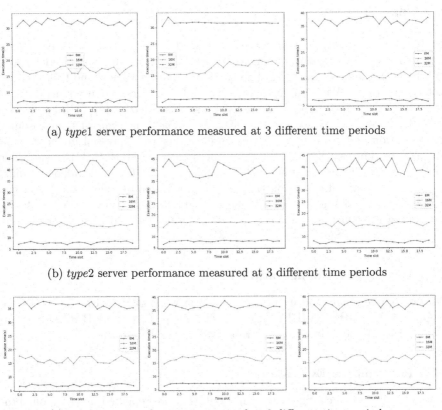

(a) *type*1 server performance measured at 3 different time periods

(b) *type*2 server performance measured at 3 different time periods

(c) *type*3 server performance measured at 3 different time periods

Fig. 4. The historical task execution time for the GaussCLegendre calculation based on different edge of servers

Table 1. The parameters used in the experiment

Parameter	Value	Meaning
ϵ	0.3	The probability of select random action
γ	0.9	The reward discount factor
lr	0.001	The learning rate of gradient descent algorithm
min_ϵ	0.05	The minimum value of ϵ
$batch\_size$	512	Sample size each step
$memory\_size$	10000	The size of samples
$\epsilon\_decrement$	0.00001	ϵ decreases each time
$replace\_target\_iter$	500	Network parameter update interval

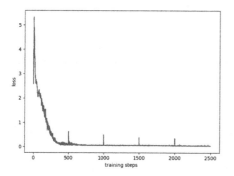

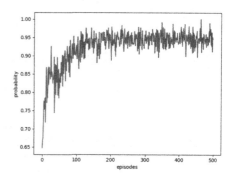

Fig. 5. The loss of evaluation network

Fig. 6. The probability of satisfying global constraint

As can be seen from Fig. 7, our method clearly outperforms baseline algorithms at all 3 time periods in terms of workflow completion time.

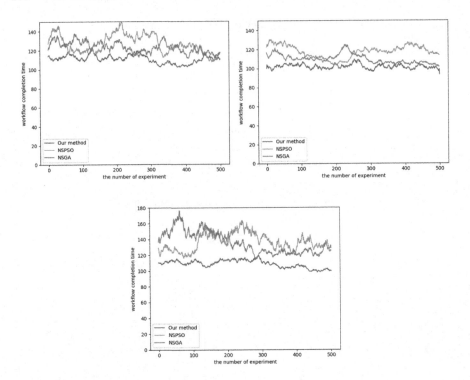

Fig. 7. Workflow completion time at 3 different time periods

6   Conclusion

In this work, a novel probability-mass function and DQN-based approach to scheduling multi-workflows upon a distributed edge-computing environment is proposed. The proposed method is capable of handling time-varying performance of edge services through probability-mass functions of historical performance data and leveraging a Deep-Q-network framework for yielding high-quality workflow scheduling plans. To validate our proposed approach, we conduct a simulative case study based on a well-known edge-service-position dataset and demonstrate that our proposed method beats its peers in terms of the scheduling performance.

Acknowledgement. This work is supported in part by Science and Technology Program of Sichuan Province under Grant 2020 JDRC0067.

References

1. Chen, X., Liu, Z., Chen, Y., Li, Z.: Mobile edge computing based task offloading and resource allocation in 5g ultra-dense networks. IEEE Access **7**, 184172–184182 (2019)

2. Ciobanu, R., Dobre, C., Balanescu, M., Suciu, G.: Data and task offloading in collaborative mobile fog-based networks. IEEE Access **7**, 104405–104422 (2019)
3. Li, G., Lin, Q., Wu, J., Zhang, Y., Yan, J.: Dynamic computation offloading based on graph partitioning in mobile edge computing. IEEE Access **7**, 185131–185139 (2019)
4. Luo, S., Wen, Y., Xu, W., Puthal, D.: Adaptive task offloading auction for industrial CPS in mobile edge computing. IEEE Access **7**, 169055–169065 (2019)
5. Zhou, J., Fan, J., Wang, J., Zhu, J.: Task offloading for social sensing applications in mobile edge computing. In: Seventh International Conference on Advanced Cloud and Big Data, CBD 2019, Suzhou, China, 21–22 September 2019, pp. 333–338. IEEE (2019)
6. Chen, H., Zhu, X., Liu, G., Pedrycz, W.: Uncertainty-aware online scheduling for real-time workflows in cloud service environment. IEEE Trans. Serv. Comput. **1**, 1 (2018)
7. Zhang, Y., Du, P.: Delay-driven computation task scheduling in multi-cell cellular edge computing systems. IEEE Access **7**, 149156–149167 (2019)
8. Cao, H., Xu, X., Liu, Q., Xue, Y., Qi, L.: Uncertainty-aware resource provisioning for workflow scheduling in edge computing environment. In: 18th IEEE International Conference On Trust, Security and Privacy in Computing and Communications / 13th IEEE International Conference On Big Data Science And Engineering, TrustCom/BigDataSE 2019, Rotorua, New Zealand, 5–8 August 2019, pp. 734–739. IEEE (2019)
9. Deng, Y., Chen, Z., Yao, X., Hassan, S., Wu, J.: Task scheduling for smart city applications based on multi-server mobile edge computing. IEEE Access **7**, 14410–14421 (2019)
10. Jian, C., Chen, J., Ping, J., Zhang, M.: An improved chaotic bat swarm scheduling learning model on edge computing. IEEE Access **7**, 58602–58610 (2019)
11. Ma, Y., et al.: A novel approach to cost-efficient scheduling of multi-workflows in the edge computing environment with the proximity constraint. In: Wen, S., Zomaya, A.Y., Yang, L.T. (eds.) Algorithms and Architectures for Parallel Processing - 19th International Conference, ICA3PP 2019, Melbourne, VIC, Australia, 9–11 December 2019, Proceedings, Part I. Volume 11944 of Lecture Notes in Computer Science, pp. 655–668. Springer, Cham (2019). https://doi.org/10.1007/978-3-030-38991-8_43
12. Peng, Q., Jiang, H., Chen, M., Liang, J., Xia, Y.: Reliability-aware and deadline-constrained workflow scheduling in mobile edge computing. In: 16th IEEE International Conference on Networking, Sensing and Control, ICNSC 2019, Banff, AB, Canada, May 9–11, 2019, pp. 236–241. IEEE (2019)
13. Bernal, J., et al.: Deep convolutional neural networks for brain image analysis on magnetic resonance imaging: a review. Artif. Intell. Med. **95**, 64–81 (2019)
14. Bouwmans, T., Javed, S., Sultana, M., Jung, S.K.: Deep neural network concepts for background subtraction: a systematic review and comparative evaluation. Neural Netw. **117**, 8–66 (2019)
15. Grekousis, G.: Artificial neural networks and deep learning in urban geography: a systematic review and meta-analysis. Comput. Environ. Urban Syst. **74**, 244–256 (2019)
16. Kaur, M., Kadam, S.: A novel multi-objective bacteria foraging optimization algorithm (MOBFOA) for multi-objective scheduling. Appl. Soft Comput. J. **66**, 183–195 (2018)

17. Zhang, L., Li, K., Li, C., Li, K.: Bi-objective workflow scheduling of the energy consumption and reliability in heterogeneous computing systems. Inf. Sci. **379**, 241–256 (2017)

18. Casas, I., Taheri, J., Ranjan, R., Wang, L., Zomaya, A.Y.: GA-ETI: an enhanced genetic algorithm for the scheduling of scientific workflows in cloud environments. J. Comput. Sci. **26**, 318–331 (2018)

19. Verma, A., Kaushal, S.: A hybrid multi-objective Particle Swarm Optimization for scientific workflow scheduling. Parallel Comput. **62**, 1–19 (2017)

20. Zhou, X., Zhang, G., Sun, J., Zhou, J., Wei, T., Hu, S.: Minimizing cost and makespan for workflow scheduling in cloud using fuzzy dominance sort based HEFT. Fut. Generation Comput. Syst. **93**, 278–289 (2019)

21. Bertsekas, D.P.: Feature-based aggregation and deep reinforcement learning: a survey and some new implementations. In: IEEE/ACM Transactions on Audio, Speech, and Language Processing, pp. 1–31 (2018)

22. Mao, H., Alizadeh, M., Menache, I., Kandula, S.: Resource management with deep reinforcement learning. In: Proceedings of the 15th ACM Workshop on Hot Topics in Networks - HotNets 2016, pp. 50–56 (2016)

23. Xue, L., Sun, C., Wunsch, D., Zhou, Y., Yu, F.: An adaptive strategy via reinforcement learning for the prisoner's dilemma game. IEEE/CAA J. Automatica Sinica **5**(1), 301–310 (2018)

24. Zhan, Y., Ammar, H.B., Taylor, M.E.: Theoretically-grounded policy advice from multiple teachers in reinforcement learning settings with applications to negative transfer. In: Proceedings of the Twenty-Fifth International Joint Conference on Artificial Intelligence. IJCAI'16, AAAI Press, pp. 2315–2321 (2016)

25. Wang, H., Huang, T., Liao, X., Abu-Rub, H., Chen, G.: Reinforcement learning for constrained energy trading games with incomplete information. IEEE Trans. Cybern. **47**(10), 3404–3416 (2017)

26. Zheng, L., Yang, J., Cai, H., Zhang, W., Wang, J., Yu, Y.: MAgent: A Many-Agent Reinforcement Learning Platform for Artificial Collective Intelligence, pp. 1–2 (2017)

27. Lowe, R., Wu, Y., Tamar, A., Harb, J., Pieter Abbeel, O., Mordatch, I.: Multi-agent actor-critic for mixed cooperative-competitive environments. In: Guyon, I., et al. (eds.) Advances in Neural Information Processing Systems 30. Curran Associates, Inc. pp. 6379–6390 (2017)

28. Duan, R., Prodan, R., Li, X.: Multi-objective game theoretic scheduling of bag-of-tasks workflows on hybrid clouds. IEEE Trans. Cloud Comput. **2**(1), 29–42 (2014)

29. Cui, D., Ke, W., Peng, Z., Zuo, J.: Multiple DAGs Workflow Scheduling Algorithm Based on Reinforcement Learning in Cloud Computing, pp. 305–311. Springer, Singapore (2016). https://doi.org/10.1007/978-981-10-0356-1_31

30. Iranpour, E., Sharifian, S.: A distributed load balancing and admission control algorithm based on Fuzzy type-2 and Game theory for large-scale SaaS cloud architectures. Future Generation Comput. Syst. **86** 81–98 (2018)

31. Jiahao, W., Zhiping, P., Delong, C., Qirui, L., Jieguang, H.: A Multi-object Optimization Cloud Workflow Scheduling Algorithm Based on Reinforcement Learning, pp. 550–559. Springer, Cham (aug (2018). https://doi.org/10.1007/978-3-319-95933-7_64

32. Guo, S., Liu, J., Yang, Y., Xiao, B., Li, Z.: Energy-efficient dynamic computation offloading and cooperative task scheduling in mobile cloud computing. IEEE Trans. Mob. Comput. **18**(2), 319–333 (2019)

33. Mnih, V., et al.: Playing atari with deep reinforcement. Learning **2055**, 1–9 (2013)

34. Lai, P., et al.: Optimal edge user allocation in edge computing with variable sized vector bin packing. CoRR abs/1904.05553 (2019)
35. Li, W., Xia, Y., Zhou, M., Sun, X., Zhu, Q.: Fluctuation-aware and predictive workflow scheduling in cost-effective infrastructure-as-a-service clouds. IEEE Access **6**, 61488–61502 (2018)
36. Beegom, A.S.A., Rajasree, M.S.: Non-dominated sorting based PSO algorithm for workflow task scheduling in cloud computing systems. J. Intell. Fuzzy Syst. **37**(5), 6801–6813 (2019)
37. Mollajafari, M., Shahhoseini, H.S.: Cost-optimized ga-based heuristic for scheduling time-constrained workflow applications in infrastructure clouds using an innovative feasibility-assured decoding mechanism. J. Inf. Sci. Eng. **32**(6), 1541–1560 (2016)

CMU: Towards Cooperative Content Caching with User Device in Mobile Edge Networks

Zhenbei Guo[1,2,5], Fuliang Li[1,2(✉)], Yuchao Zhang[3], Changsheng Zhang[1], Tian Pan[3,4], Weichao Li[2,6(✉)], and Yi Wang[2,6(✉)]

[1] Northeastern University, Shenyang, People's Republic of China
lifuliang207@126.com
[2] Peng Cheng Laboratory, Shenzhen, People's Republic of China
liweichao@gmail.com, wy@ieee.org
[3] Beijing University of Posts and Telecommunications,
Beijing, People's Republic of China
[4] Purple Mountain Laboratories, Nanjing, People's Republic of China
[5] Key Laboratory of Computer Network and Information Integration
(Southeast University), Ministry of Education, Nanjing, People's Republic of China
[6] Institute of Future Networks, Southern University of Science and Technology,
Shenzhen, People's Republic of China

Abstract. Content caching in mobile edge networks has stirred up tremendous research attention. However, most existing studies focus on predicting content popularity in mobile edge servers (MESs). In addition, they overlook how the content is cached, especially how to cache the content with user devices. In this paper, we propose CMU, a three-layer (Cloud-MES-Users) content caching framework and investigate the performance of different caching strategies under this framework. A user device who has cached the content can offer the content sharing service to other user devices through device-to-device communication. In addition, we prove that optimizing the transmission performance of CMU is an NP-hard problem. We provide a solution to solve this problem and describe how to calculate the number of distributed caching nodes under different parameters, including time, energy and storage. Finally, we evaluate CMU through a numerical analysis. Experiment results show that content caching with user devices could reduce the requests to Cloud and MESs, and decrease the content delivery time as well.

Keywords: Mobile edge networks · Cooperative caching · Device-to-device communication · Content delivery

1 Introduction

To cope with massive data traffic brought by the ever-increasing mobile users, the traditional centralized network model exhibits the disadvantages of high

© Springer Nature Switzerland AG 2020
W.-S. Ku et al. (Eds.): ICWS 2020, LNCS 12406, pp. 210–227, 2020.
https://doi.org/10.1007/978-3-030-59618-7_14

delay, poor real-time and high energy consumption. To solve these problems, the mobile edge network model is proposed, which can provide cloud computing and caching capabilities at the edge of cellular networks. According to the survey report by Cisco [1], the number of connected devices and connections will reach 28.5 billion by 2022, especially mobile devices, such as smartphones, which will reach 12.3 billion. Cisco also points out that traffic of video will account for 82% of the total IP traffic. Due to the massive content transmission traffic generated by user requests, content caching is regarded as a research hotspot of mobile edge network [2,3]. Liu et al. [4] also indicates that mobile edge caching can efficiently reduce the backhaul capacity requirement by 35%.

When a user requests the content in a traditional centralized network, the content will be provided by a remote server or Cloud, wherein there is usually duplicated traffic during the content transmission. By caching the content from the Cloud to the edge of the network (e.g., gateway and base station), the duplicated traffic can be avoided when the user chooses the closest mobile edge server. At the same time, it has a better network quality than the traditional centralized network. Due to the limited storage space, user devices cannot cache all the contents. But with the development of the hardware, the computing and storage capabilities of mobile devices have been improved greatly. Even though the storage capabilities of a user device cannot be compared with that of Cloud and MES, we can build a huge local content caching network relying on the explosive growth of mobile devices, which could use D2D to provide content sharing services [13,14]. More and more studies show that a local content caching network has a great potential to achieve content sharing.

In this paper, we apply a three-layer content caching framework in the mobile edge network and aim to investigate the performance of different caching ways under this framework. Caching the content from Cloud and MES to user device could reduce the requests to Cloud and MES, as well as save the valuable bandwidth resources. It can also reduce the content transmission time and energy consumption. Paakkonen et al. [18] studies the performance of local content caching network when using different caching ways. We refer their caching strategies and make these caching strategies applicable to both MESs and user devices. We assume that the content popularity is known and it conforms to the ZipF model [22]. The main contributions of this paper can be summarized as follows. (1) We combine MESs and user devices to cache the content to create a multilayer mobile edge caching framework. To the best of our knowledge, content caching that is assisted by user devices in mobile edge network has not been well studied in previous work. (2) We prove that minimizing transmission costs between MESs and user devices is an NP-Hard problem. We provide a solution to solve this problem and describe how to calculate the number of distributed caching nodes in a cluster. (3) We evaluate the performance of different caching strategies under the proposed framework through a numerical analysis. Results show that caching contents with user devices is feasible and effective.

The rest of the paper is organized as follows. Section 2 presents the related work. Section 3 introduces the proposed framework. Problem description and theoretical derivation are presented in Sect. 4. Numerical analysis results are shown in Sect. 5. Finally, the paper is concluded in Sect. 6.

2   Related Work

Recent studies use different machine learning methods to predict the popular content or apply other technologies to maximize the storage and computing capabilities of MESs [5–12]. For example, Chen et al. [5] used a self-supervising deep neural network to train the data and predict the distribution of contents. Liu et al. [9] proposed a novel MES-enabled blockchain framework, wherein mobile miners utilize the closest MES to compute or cache the content (storing the cryptographic hashs of blocks). In these studies, even there are many caching strategies, the predicted contents are simply copied in MESs and the main optimization objective is the MES. A survey of content caching for mobile edge network [3] indicated that mobile traffic could reduce one to two thirds by caching at the edge of the network. In addition, D2D as one of the key technologies of 5G could easily help user to utilize the locally stored resources. And the QoE of users could have a great improvement by local content caching [15,17]. Video traffic accounts for the vast majority of IP traffic [1], and Wu et al. [16] proposed a user-centric video transmission mechanism based on D2D communications allowing mobile users to cache and share videos with each other. Paakkonen et al. [18] also investigated different D2D caching strategies.

Existing studies have combined the local devices with the cellular network for traffic offloading. Kai et al. [23] considered a D2D-assisted MES scenario that achieves the task offloading. Zhang et al. [24] focused on the requested contents cached in the nearby peer mobile devices. They aimed to optimize the D2D throughput while guaranteeing the quality of D2D channels. However, devices randomly cached the popular contents and there was no global content caching policy. In addition, devices are distributed within a small range so that the communication could be established at any time. But in practice, we should consider the scenario where the requesting device is far from the caching device.

3   System Model

The mobile edge network structure used in this paper is shown in Fig. 1. The top layer is the Cloud, followed by the mobile edge server layer and the local user device layer. In the local user device layer, the devices (nodes) are divided into two types, i.e., Normal node and Caching node. When a device requests the content, the sequence of the request is Local >> MES >> Cloud.

3.1   Caching Strategies

Four caching strategies are adopted, and some contents are randomly distributed across the MESs in the beginning.

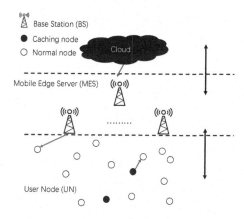

Fig. 1. The architecture of the mobile edge network.

Default Caching (DC): Some contents are randomly stored in the MESs and remain unchanged. If the MESs do not have the requested content, the request will be sent to the Cloud.

Replication Caching (RC): User caching nodes are called the caching nodes, and MES is called the caching server. The MESs and user nodes store a full copy of each content when the transmission process is finished. User nodes are regarded as the content servers after they cache the contents. They could send the desired content to the requesting device by the D2D communication. But if the caching node leaves, all the caching contents are lost. It has to request the desired content from the farther caching nodes or the MESs.

Distributed Replication Caching (DRC): User caching nodes are called the distributed caching nodes, and MES is called the distributed caching server. Assuming that the number of user nodes is N, and n ($n \ll N$) nodes are selected as the distributed caching nodes by MESs. Distributed caching nodes are respon- sible for content sharing to the normal nodes. Note that, distributed caching nodes do not actively request any content for themselves. If a distributed caching node requests the content for a normal node from the upper layer, all distributed caching nodes cache the full copy of each content, which is returned from the upper layer. That is to say each content is cached in n nodes. When a distributed caching node leaves its cluster, an available node from that cluster will replace it. Using this strategy, the cached contents are not easily lost. But there is an addi- tional compensation consumption when a new distributed caching node recovers the previously cached content.

Distributed Fragment Caching (DFC): Different from DRC, the content is divided into several parts according to the number of distributed caching nodes in the requesting cluster. That is to say each distributed caching node stores a part of the content. The upper layer performs the content fragmentation automatically during the experiment. Note that MESs use the DRC caching strategy when user nodes use DFC.

3.2   User State

We assume that state space of a user is $State = \{Leave, Stay, Request\}$, and the state-transition is a Markov chain with two reflective walls on a straight line as shown in Fig. 2.

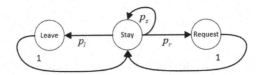

Fig. 2. The state-transition diagram.

Where *Leave* means the user leaves and deletes the cached contents. The probability of *Leave* is p_l. We assume that a user returns back to the cluster conforming to the Poisson process. *Stay* means the user stay in the place with no actions, and the probability of *Stay* is p_s. *Request* means the user requests the desired content, and the probability of *Request* is p_r. We have $p_s + p_r + p_l = 1$. We assume that the number of all the users is $\mathcal{N}$, $\mathcal{N} = \{1, 2, ..., n\}$. The state matrix of the user is denoted by $State = \{State_{ij} \mid i \in N, j = 1, 2, 3\}$. We use 0 and 1 to represent the user state, where 1 represents that the user is in the corresponding state. The users can utilize the Service Set Identifier (SSID) to access the cached content and sense the state of other users. In addition, the users could receive the data from the MESs while they share the data to the other user devices. To avoid the transmission game, we define a game strategy set of $a_n \in \{0\} \cup \mathcal{N}$, wherein $a_n > 0$ means the user of n is willing to provide the content sharing service. The value of a_n is subtracted by 1 whenever the content is transmitted successfully. If $a_n = 0$, the user n is willing to request its desired content, and set a_n with the default value after successfully requesting.

4   Theoretical Analysis

We first introduce the NP-Hard problem and present how to calculate the performance of each caching strategy. Then, the derivation of calculating the number of the distributed caching nodes is presented.

4.1   Problem Formulation and Solution

The contents are denoted as $\mathcal{R} = \{r_1, ..., r_E\}$, and the size of each content is set to s. There are m MESs and the set of MESs is denoted as $\mathcal{S} = \{S_1, ..., S_m\}$. The set of $\alpha = \{\alpha_1, ..., \alpha_m\}$ means the maximum storage space of the MESs, while $\beta = \{\beta_1, ..., \beta_n\}$ is the maximum storage space of the users. And we have $\beta << \alpha$. The content caching matrix of the MESs is $X = \{X_{m,E} \mid S_m \in \mathcal{S}, r_E \in \mathcal{R}\}$, and the content matrix of the users is $Y = \{Y_{n,E} \mid n \in \mathcal{N}, r_E \in \mathcal{R}\}$. If a mobile edge server of m has cached the content of E, then $X_{m,E} = 1$, otherwise $X_{m,E} = 0$. The content matrix of the users has the same operations. The storage space of the MESs and the users should meet the following constraints.

$$\begin{cases} \sum_{i=1}^{E} X_{m,i} \cdot s \leq \alpha_m, S_m \in \mathcal{S} \\ \sum_{i=1}^{E} Y_{n,i} \cdot s \leq \beta_n, n \in \mathcal{N} \end{cases} \tag{1}$$

When the storage space reaches to the maximum capacity and a copy of new content needs to be cached, the Least Recently Used (LRU) algorithm is utilized to realize the space release and content replace. In addition, we assume that the transmission consumption of the users is η_u, and the transmission rate is TR_u. The values for each MES are marked as η_s and TR_s. Similarly, the symbols of the Cloud are η_c and TR_c. We assume that C^E is the energy consumption and T^E is the time overhead caused by transmitting the content of E. We should make sure that the energy consumption and the time overhead are minimized during each transmission. e.g., node k transmits the content of E to node o, $C_{k,o}^E$ and $T_{k,o}^E$ should reach the minimum. That is to say the distance or the route path between them should be minimized. The distance matrix of the users is denoted as $DU = \{DU_{i,j} \mid i,j = 1,2,...,n; i \neq j\}$, while for each MES, the expression is $DS = \{DS_{i,j} \mid i,j = 1,2,...,m; i \neq j\}$. When the content of E is transmitted from node k to node o, the following constraints (Problem 1, P1) should be satisfied.

$$(P1) \quad Min \quad (C^E + T^E)$$
$$s.t. \quad \sum_{i=1}^{E} X_{k,i} \cdot s \leq \alpha_k, S_k \in S$$
$$\sum_{i=1}^{E} Y_{k,i} \cdot s \leq \beta_k, k \in N$$
$$DS_{k,o} = \underset{g \neq o}{Min} \{DS_{g,o} \mid g = 1,2,...,m; X_{g,E} = 1\}$$
$$DU_{k,o} = \underset{g \neq o}{Min} \{DU_{g,o} \mid g = 1,2,...,n; Y_{g,E} = 1\}$$

Where, after the transmission is completed, the required caching space should not exceed the maximum storage capacity of the node. During the transmission process, the shortest distance or route path should be chosen.

Theorem 1. *Energy consumption and time overhead reach the minimum in P1 is an NP-Hard problem.*

Proof. Travel Sale-man Problem (TSP) is a classical NP-Hard problem. During the experiment, we assume that any object (i.e., any user) of $o \in \{\mathcal{N}\}$ requests the content from the Cloud. The Cloud sends the content to the request object of o going through the MES layer and the user layer. The route path of the user layer is R_{user}, and the consumption of the single-hop routing is ν_u. Let R_{sever} denote the route path of the MES layer, and the consumption of the single-hop routing is ν_s. We have $\nu_{all} = \nu_u * R_{user} + \nu_s * R_{sever}$, and our goal is to minimize ν_{all}. We treat the request object of o as the Sale-man and consider the Cloud as the destination. The total number of the travel itinerary between them is $R_{user} + R_{sever}$, and there are $R_{user}^{R_{sever}}$ schemes. We need to find the smallest ν_{all} of these schemes. Therefore, P1 is equivalent to the Travel Sale-man Problem, i.e., P1 is NP-Hard.

Since P1 is NP-hard, we need to resort to a sub-optimal solution for P1. We use a clustering algorithm to reduce the number of the available route paths, e.g., the improved K-means++ algorithm. In RC, the requesting nodes give priority to scanning other nodes within their own clusters. Frequent communication can enrich the cache community greatly. In the distributed caching strategies, each cluster has its own distributed caching nodes, wherein the normal nodes could request the contents from the closest distributed caching nodes.

The clustering algorithms can be used for both the MES layer and the user layer, and we take the user layer as an example to illustrate. The coordinate vectors of the users is regarded as the training sample, which is denoted as $I = \{I_i \mid I_i \in R^2, i = 1, 2, 3, ..., n\}$, e.g., $I_i = \{x_i, y_i\}$. There are K clusters, e.g., $\Phi = \{\Phi_1, ..., \Phi_K\}$. The center of the cluster is denoted as $\pi = \{\pi_i, ..., \pi_K\}$. The number of the cluster members is denoted as $\varphi = \{\varphi_1, ..., \varphi_K\}$. We calculate the euclidean distance of each sample from the center. For $\forall I_i \in \Phi_i$, we have:

$$\|I_i - \pi_i\|_2 = \min_j \|I_i - \pi_j\|_2. \tag{2}$$

The indicator function g is used to check whether the distance between the center and each sample is the shortest, which is expressed as

$$g_{i,j} = \begin{cases} 1, & \|I_i - \pi_i\|_2 = \min_j \|I_i - \pi_j\|_2 \\ 0, & Otherwise. \end{cases} \tag{3}$$

When $\sum_{i=1}^{n} \sum_{j=1}^{K} g_{i,j}$ is at its maximum, the error rate of cluster is the lowest. The new centres will be calculated by the following expression.

$$\pi_j = \frac{\sum_{i=1}^{n} g_{i,j} I_i}{\sum_{i=1}^{n} g_{i,j}}. \tag{4}$$

We use the improved K-means++ algorithm as shown in Algorithm 1.

If the center nodes are randomly selected and they are too close to each other, the convergence will become slow, which degrades the performance of K-means.

Therefore, how to select the initial center nodes is optimized in Algorithm 1 (line 1–line 11). We select the farthest node from the existing centres every time, so the selected center could cover all the samples as far as possible. When the center selection process is completed, the iteration process is enabled (line 12–line 17).

4.2   Performance Formulation

To better describe the performance, we have $\mathcal{B} = \{b_1, ..., b_n, b_{n+1}, ..., b_{n+m}\}$. For example, $b_u^E = 1$ is utilized to represent that the content of E is transmitted by the user node of u.

Algorithm 1. K-means++

Input: I, K
Output: Φ
1: Initialize the number of cluster center $k = 0$, the number of iteration $n = 0$, distance array $D = \{0\}$, cluster center array $\pi = \{0\}$;
2: Let $k = 1$, a node $\forall I_k \in I$ is randomly selected as the first center, update π;
3: **while** $k < K$ **do**
4:     $D = \{0\}$;
5:     **for** each non-central node $\forall I_i \in I$ **do**
6:         Calculate the shortest distance by equation (2);
7:         Update D;
8:     **end for**
9:     Select the node with the maximum value in D as the new center;
10:    Update k and π;
11: **end while**
12: Let the number of iteration $n = 1$.
13: **repeat**
14:    Partition each non-central node from I by equation (3).
15:    Calculate the new centres by equation (4) and update π ;
16: **until** $\pi_j(n+1) == \pi_j(n), \forall j = 1, 2, ..., K$

Default Caching and **Replication Caching** have the same performance expressions from the aspects of storage, energy and time. The utilized storage space is represented as follows.

$$\left(\sum_{i=1}^{m}\sum_{j=1}^{E} X_{i,j} + \sum_{i=1}^{n}\sum_{j=1}^{E} Y_{i,j}\right) \cdot s. \tag{5}$$

C^E is expressed as:

$$C^E = \begin{cases} \eta_u \cdot s & b_u^E = 1 \\ \eta_s \cdot s & b_u^E = 0, b_s^E = 1 \\ (\eta_s + \eta_c) \cdot s & b_u^E = 0, b_s^E = 0 \end{cases} \tag{6}$$

T^E is expressed as:

$$T^E = \begin{cases} \frac{s}{TR_u} & b_u^E = 1 \\ \frac{s}{TR_s} & b_u^E = 0, b_s^E = 1 \\ (\frac{s}{TR_s} + \frac{s}{TR_c}) & b_u^E = 0, b_s^E = 0 \end{cases} \tag{7}$$

Distributed Caching. When using the distributed caching strategies of DRC and DFC, the requesting nodes first request a distributed caching node. If the required content is not found, the distributed caching node will request the closest distributed caching server. Then, the required content is cached by the distributed caching node and transmitted to the requesting nodes by D2D communication. To prevent the requesting nodes from the same cluster to occupy too much bandwidth resources of the closest distributed caching server, contents are always sent from distributed caching servers to distributed caching nodes by default. If the distributed caching server does not have the requested content, it will request the Cloud with the same process. There are K clusters in the user layer divided by K-means++. However, each cluster cannot specify only one node as the distributed caching node. If there is only one distributed caching node, it will be difficult to handle a large number of concurrent requests, because transmitting the content to the requesting nodes from one distributed caching node by D2D communication is infeasible.

Taking the user cluster as an example, ξ denotes the number of distributed caching nodes. It is assumed that ξ_K nodes are selected from cluster K as the distributed caching nodes to cache and share the contents. For the distributed caching strategies of DRC and DFC, the multi-objective optimization problem (Problem 2, P2) is expressed as follows.

$$P2: \quad min \quad f_1 = \sum_{i=1}^{\xi_K} \sum_{j=1}^{E} Y_{i,j} \cdot s$$

$$min \quad f_2 = \left\lceil \frac{\varphi_K - \xi_K}{\xi_K} \right\rceil \cdot T$$

$$min \quad f_3 = \sum_{i=1}^{\varphi_K - \xi_K} \cdot C_i$$

$$s.t. \quad 0 < \xi_K < \varphi_K,$$

$$\sum_{i=1}^{E} Y_{k,i} \cdot s \le \beta_k, \quad k \in \xi_K.$$

Where, f_1 denotes the storage space, and f_1 of each node can not exceed the limit value. Similarly, f_2 denotes the time required to complete all of the requests, and f_3 denotes the energy consumption caused by all of the successful requests. The optimization object of f_1 is opposite to f_2 and f_3. When f_1 decreases, fewer distributed caching nodes provide the content sharing services, but at the same time, f_2 and f_3 increase. When minimizing f_2 and f_3, f_1 increases.

In order to solve P2, the Linear Weighting Method is adopted to make P2 become a single-objective optimization problem. The weight vectors are denoted as $u = \{u_i \mid u_i \geq 0, i = 1, 2, 3\}$, and $\sum\limits_{i=1}^{3} u_i = 1$. Each weight coefficient is corresponding to f_i in P2. Then P2 can be expressed as follows.

$$\min_{\xi_K} \sum_{i=1}^{3} u_i f_i(\xi_K). \tag{8}$$

Let $u_i = \xi_i$. And ξ_i is satisfied by expression (9).

$$f_i(\xi_i) = \min_{\xi_k \in \varphi_K} f_i(\xi_k) \qquad i = 1, 2, 3. \tag{9}$$

Where, ξ_i is the optimization value making the object of f_i reach the minimum. Then we have the following expression.

$$u = [1, \varphi_K - 1, \varphi_K - 1]^T. \tag{10}$$

We substitute Eq. (10) into Eq. (8).

$$\min \quad F(\xi_K) = \frac{\bar{\beta}}{2\varphi_K - 1} \cdot \xi_K + \frac{(\varphi_K - 1) \cdot \bar{T}}{2\varphi_K - 1} \cdot \frac{\varphi_K - \xi_K}{\xi_K}$$
$$+ \frac{(\varphi_K - 1) \cdot \bar{C}}{2\varphi_K - 1} \cdot (\varphi_K - \xi_K). \tag{11}$$

$\bar{\beta}$ is the average caching space, $\bar{T}$ is the average transmission time and $\bar{C}$ is the average energy consumption. We take the derivative of ξ_K and set Eq. (11) to 0. Then we have the following expression.

$$\frac{\bar{\beta}}{2\varphi_K - 1} - \frac{(\varphi_K - 1) \cdot \bar{T}}{2\varphi_K - 1} \cdot \frac{\varphi_K}{\xi_K^2} - \frac{\varphi_K - 1}{2\varphi_K - 1} \cdot \bar{C} = 0$$

$$\bar{\beta} - \frac{(\varphi_K^2 - \varphi_K) \cdot \bar{T}}{\xi_K^2} - (\varphi_K - 1) \cdot \bar{C} = 0$$

$$\frac{(\varphi_K^2 - \varphi_K) \cdot \bar{T}}{\xi_K^2} = \bar{\beta} - (\varphi_K - 1) \cdot \bar{C}$$

$$\xi_K = \sqrt{\left| \frac{(\varphi_K^2 - \varphi_K) \cdot \bar{T}}{\bar{\beta} - (\varphi_K - 1) \cdot \bar{C}} \right|}.$$

The result is an efficient solution to Eq. (11). It is round up to an integer and considered as the number of the distributed caching nodes. We then discuss the performance of DRC and DFC.

1) **Distributed Replication Caching (DRC):** MES layer has KS clusters, e.g., $\psi = \{\psi_1, ..., \psi_{KS}\}$. The number of the distributed caching servers in each

---

Algorithm 2. Adding and deleting the content

Input: r_E, S, α, X
Output: X
1: Initialize the content $r = null$;
2: **for** i=0 to the number of the distributed caching servers **do**
3:    **while** the caching space of $S_i >= \alpha_i$ **do**
4:       Find r by LRU;
5:       Delete r and update X.
6:    **end while**
7: **end for**
8: **for** i=0 to the number of the distributed caching servers **do**
9:    **if** $X_{i,E} == 0$ **then**
10:       Cache r_E and update $X_{i,E} = 1$.
11:    **end if**
12: **end for**

---

cluster is denoted as $\bar{S}$. The utilized caching space can be estimated by the following expression.

$$\sum_{i=1}^{KS}\sum_{j=1}^{E} X_{m,j} \cdot s \cdot \bar{S}_i + \sum_{k=1}^{K}\sum_{j=1}^{E} Y_{n,j} \cdot s \cdot \xi_k \qquad (12)$$
$$S_m \in \psi_i, \quad n \in \Phi_k.$$

The distributed caching servers within a cluster cache the same contents so that only the copies of the content cached in one distributed caching server need to be recorded. Then, multiplying the copies by the content size and the number of the distributed caching servers, the result is the utilized caching space. In addition, in order to guarantee the consistency of the distributed caching servers, Algorithm 2 is applied to add and delete the content. We take the MES layer as an example to illustrate the algorithm. The energy consumption can be calculated by Eq. (6). But if the content of E is transmitted from the upper layer, it needs to be cached in the distributed caching server, as well as the distributed caching nodes. So each layer will have additional transmission consumption: multiplying the number of distributed caching servers or nodes by C^E. When a distributed caching node leaves, an available node from the same cluster will be selected as the new distributed caching node. The selection rule follows Algorithm 1 (line 1– line 11). The extra consumption caused by recovering the cached content is also calculated by Eq. (6). We denote $\bar{N}_r$ as the number of nodes requesting content of E simultaneously. If the requesting nodes are more than the distributed caching nodes, some requesting nodes have to wait until the distributed caching nodes have the capacity to provide the content sharing services. The required time is calculated by the following expression.

$$T^E = \begin{cases} \frac{s}{TR_u}\left\lceil\frac{\bar{N}_r}{\xi}\right\rceil & b_u^E = 1 \\ \frac{s}{TR_u}\left\lceil\frac{\bar{N}_r}{\xi}\right\rceil + \frac{s}{TR_s} & b_u^E = 0, b_s^E = 1 \\ \frac{s}{TR_u}\left\lceil\frac{\bar{N}_r}{\xi}\right\rceil + \frac{s}{TR_s} + \frac{s}{TR_c} & b_u^E = 0, b_s^E = 0. \end{cases} \tag{13}$$

2) **Distributed Fragment Caching (DFC):** The content is divided into several parts. The size of each part of the content is different across the clusters. The caching space is expressed as follows.

$$\sum_{i=1}^{KS}\sum_{j=1}^{E} X_{m,j}\cdot s + \sum_{k=1}^{K}\sum_{j=1}^{E} Y_{n,j}\cdot s \tag{14}$$
$$S_m \in \psi_i, \quad n \in \Phi_k.$$

If the target distributed caching node is busy, other parts of the required content will be transmitted from other distributed caching nodes. The required time is expressed as follows.

$$T^E = \begin{cases} \frac{s}{TR_u}\left\lceil\frac{\bar{N}_r}{\xi}\right\rceil & b_u^E = 1 \\ \frac{s}{TR_u}\left\lceil\frac{\bar{N}_r}{\xi}\right\rceil + \frac{s}{TR_s\cdot\xi} & b_u^E = 0, b_s^E = 1 \\ \frac{s}{TR_u}\left\lceil\frac{\bar{N}_r}{\xi}\right\rceil + \frac{s}{TR_s\cdot\xi} + \frac{s}{TR_c} & b_u^E = 0, b_s^E = 0. \end{cases} \tag{15}$$

The energy consumption can be calculated by Eq. (6). There is no extra transmission consumption when the content is transmitted from the upper layer, because the total size that needs to be transmitted is equal to the size of the original content. When a node is selected as the new distributed caching node, the extra consumption caused by recovering the cached content is calculated by $\sum_{i=1}^{E} C^i$, wherein

$$C^i = \begin{cases} \eta_u \cdot \frac{s}{\xi} & if \quad Y_{i,} = 1 \\ 0 & Otherwise. \end{cases} \tag{16}$$

If all the distributed caching nodes are busy, the MESs will help to recover the cached content. Then, for each content that needs to be recovered, the extra consumption is calculated by $C = \eta_s \cdot \frac{s}{\xi}$.

5   Experiment

5.1   Experiment Setup

10 MESs and 100 user nodes are randomly distributed within $500 \times 500\,\text{m}$. The simulation ends after the transmission is performed for 5000 times successfully. The caching space of each MES is 100 MB, while the size of each user node is 20 MB. We assume that the expectation of the Poisson process for the leaving nodes is 100. All the MESs are distributed caching servers in their own clusters, and they will not leave their places. The transmission parameters are set

according to a survey report [25], where η_u is 5 j/MB. The MESs and the Cloud utilize the cellular network, where η_s and η_c are 100 j/MB. The transmission rate of the user is between 54 Mb/s and 11 Mb/s within 200 m, while for the MESs, the value is between 1 Mb/s and 0.5 Mb/s within 500 m. The value for the Cloud is set to 0.5 Mb/s. There are 1000 video items and the size of each item is 20 Mb. The popularity distribution of the items is generally modeled as a ZipF distribution. The default shape parameter θ is set to 0.56 [26]. The default game strategy value of a is set to 1. The default requesting probability of p_r is 0.6, and p_s is equal to p_l, i.e., 0.2.

In addition, we add two comparison objects, i.e., DRC-Extra and DFC-Extra. All user nodes in RC could cache the contents, while only the distributed caching nodes have the caching ability in DRC and DFC. Therefore, to guarantee the same caching capacity, we increase the caching space of the distributed caching nodes in DRC-Extra and DFC-Extra. The new caching space of each distributed caching node is calculated as follows.

$$new\ caching\ space = \frac{the\ total\ caching\ space\ of\ RC}{the\ number\ of\ the\ distributed\ caching\ nodes}. \quad (17)$$

We calculate the **average transmission time (ATT)**, **average energy consumption (AEC)** and **the number of local requests (LR)** that are served by the caching nodes or the distributed caching nodes. These metrics are used to evaluate the effect on reducing the requests to the MESs and the Cloud. The **caching space (CS)** is adopted for choosing an appropriate caching strategy when the caching space is restricted.

5.2 Numerical Analysis

The DC strategy is utilized as a baseline for numerical analysis. In DC, contents are randomly stored across the MESs and the cached contents remain unchanged during the whole experiment. The numerical analysis results under different parameters are shown in Fig. 3a to Fig. 3o.

In Fig. 3a to Fig. 3d, *items* and θ are set to 1000 and 0.56 respectively, while p_r changes from 0.1 to 0.9. The increase of the request probability means fewer nodes leave away and more different contents are cached. As a result, the caching space increases (Seen in Fig. 3c), while the energy consumption decreases (Seen in Fig. 3b). For DFC, DFC-Extra and RC, the copies of local contents increase with the growth of p_r. So the transmission time of them decreases. For DRC and DRC-Extra, since they have to cache the full copy of each content and the caching space is limited, the amount of the content cached locally is small. They have to request new contents from the upper layer frequently, so the transmission time of DRC and DRC-extra remains unchanged. Since the amount of content cached on MESs increases, DRC and DRC-Extra spend less time than DC. Note that the time consumption of DFC-Extra is high when p_r is equal to 0.1. The low request rate causes a large number of distributed caching nodes to leave, and the requesting nodes have to wait for new distributed caching nodes to

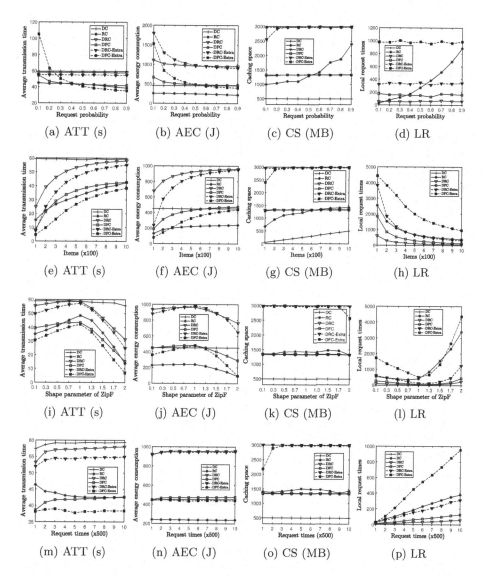

Fig. 3. Experimental results under different parameters. a to d, *items* = 1000, θ = 0.56, $p_r \in [0.1, 0.9]$. e to h, θ = 0.56, p_r = 0.6, *items* $\in [100, 1000]$. i to l, p_r = 0.6, *items* = 1000, $\theta \in [0.1, 2]$. m to p, p_r = 0.6, θ = 0.56, *items* = 1000.

recover the cached content. Compared with RC, cached contents are not lost when using distributed caching strategies so that the caching space remains unchanged. Meanwhile, the extra consumption of recovering the cached contents causes higher energy consumption when using distributed caching strategies.

In Fig. 3e to Fig. 3h, θ and p_r are set to 0.56 and 0.6 respectively, while *items* change from 100 to 1000. When the number of items is 100 (most of the contents are cached locally), the performance of each strategy is optimal, especially DRC and DRC-Extra. It means that the distributed caching strategies perform better with a larger caching space. The increase of items means more different contents can be requested and the content caching update is more frequent, which causes the increase of the transmission time (Seen in Fig. 3e), the energy consumption (Seen in Fig. 3f) and the caching space (Seen in Fig. 3g). Meanwhile, it reduces the number of local requests (Seen in Fig. 3h). The caching space of distributed caching strategies remains unchanged because the number of the distributed servers is fixed and the cached contents are not lost. The caching space of RC fluctuates greatly caused by the leaving caching nodes. DRC and DRC-Extra have to cache several full backups after each successful request, resulting in the highest energy consumption (Seen in Fig. 3f).

In Fig. 3i to Fig. 3l, $p_r = 0.6$ and $items$ are set to 0.6 and 1000 respectively, while θ changes from 0.1 to 2. The parameter of θ determines the popularity of the items. When θ is too small or too large, some contents will be requested frequently. The performance of each strategy is optimal when θ is equal to 2, where the popularity of the items is high. When θ changes from 0.1 to 1, the transmission time increases (Seen in Fig. 3i), wherein the popularity of items are evenly distributed gradually. The cached contents update frequently, resulting in more energy consumption (Seen in Fig. 3j) and fewer local requests (Seen in Fig. 3l). When θ is greater than 1, some contents are requested frequently. The transmission time and energy consumption decrease, while the number of local requests increases. The caching space (Seen in Fig. 3k) remains high because other requested contents are still cached.

In Fig. 3m to Fig. 3p, p_r, $items$ and θ are set to 0.6, 1000 and 0.56 respectively. We evaluate the performance under different successful requests. As depicted in Fig. 3m, DFC-Extra has the minimum transmission time, and the transmission time of RC and DFC is gradually approaching to each other. In the beginning, the amount of the cached content of RC is small. However, the cached content of RC is enriched gradually and the transmission time decreases, while the transmission time of the distributed caching strategies increases due to the small amount of the cached content and frequent content caching updates. The energy consumption increases slightly (Seen in Fig. 3m) caused by updating and recovering the content. The caching space (Seen in Fig. 3o) of the distributed caching strategies remains unchanged after it reaches the maximum limit, while the caching space of RC still fluctuates. The local requests (Seen in Fig. 3p) of DFC-Extra and RC tend to grow faster due to caching a large number of popular contents.

The findings are summarized as follows. (1) The transmission time and energy consumption of RC, DFC and DFC-Extra are better than other caching strategies in most cases. DFC-Extra performs best, but it needs a large cache space and presents high energy consumption as well. (2) RC performs well in all the scenarios. However, the cached contents are easily lost and the performance degrades if the nodes in the community are not willing to share the contents. (3) All the distributed caching strategies work badly when the nodes leave frequently (e.g.,

$p_r = 0.1$). However, all of them work well under high request probability and popularity (e.g., $\theta = 2$). The distributed caching strategies need more caching space to work better, but they also cause higher energy consumption. DRC and DRC-Extra seem to present the worst performance, but they can keep working when only one distributed caching node provides service. In DFC and DFC-Extra, the nodes have to wait when one of the distributed caching nodes is lost. (4) More importantly, using user devices to cache the content could not only reduce the requests to the Cloud and the MESs, but also decrease the content transmission time as shown in Fig. 3a and Fig. 3d.

6   Conclusions

In this paper, we propose CMU, a three-layer (Cloud-MESs-Users) content caching framework, which could offload the traffic from Cloud-MESs to user devices. We describe the content caching problem and evaluate the performance of different caching strategies under this framework. Numerical results show that content caching with user devices could reduce the requests to Cloud-MESs, as well as decrease the content transmission time. To attract more nodes to participate in content caching and sharing, an incentive mechanism is needed for the distributed caching strategies in the future.

Acknowledgment. The work is supported by the National Key Research and Development Program of China under Grant Nos. 2018YFB1702000, 2018YFB1800201 and 2019YFB1802600; the National Natural Science Foundation of China under Grant Nos. 61602105, 61702049 and 61872420; the project of PCL Future Greater-Bay Area Network Facilities for Large-scale Experiments and Applications under Grant No. LZC0019; the Guangdong Key Research and Development Program under Grant No. 2019B121204009; the Fundamental Research Funds for the Central Universities under Grant Nos. N2016005, N171604006 and 2019RC03; CERNET Innovation Project under Grant Nos. NGII20170121 and NGII20180101.

References

1. Cisco: Cisco visual networking index: global mobile data traffic forecast update, 2017–2022 (2019)
2. Wang, X., Chen, M., Taleb, T., Ksentini, A., Leung, V.C.M.: Cache in the air: exploiting content caching and delivery techniques for 5G systems. IEEE Commun. Mag. **52**(2), 131–139 (2014)
3. Wang, S., Zhang, X., Zhang, Y., Wang, L., Yang, J., Wang, W.: A survey on mobile edge networks: convergence of computing, caching and communications. IEEE Access **5**, 6757–6779 (2017)
4. Liu, D., Yang, C.: Energy efficiency of downlink networks with caching at base stations. IEEE J. Sel. Areas Commun. **34**(4), 907–922 (2016)
5. Chen, J., Chen, S., Wang, Q., Cao, B., Feng, G., Hu, J.: iRAF: a deep reinforcement learning approach for collaborative mobile edge computing IoT networks. IEEE Internet Things J. **6**(4), 7011–7024 (2019)

6. Jiang, W., Feng, G., Qin, S., Liang, Y.: Learning-based cooperative content caching policy for mobile edge computing. In: 2019 IEEE International Conference on Communications, ICC 2019, Shanghai, China, 20–24 May 2019, pp. 1–6 (2019)
7. Ale, L., Zhang, N., Wu, H., Chen, D., Han, T.: Online proactive caching in mobile edge computing using bidirectional deep recurrent neural network. IEEE Internet Things J. **6**(3), 5520–5530 (2019)
8. Cui, Q., et al.: Stochastic online learning for mobile edge computing: learning from changes. IEEE Commun. Mag. **57**(3), 63–69 (2019)
9. Liu, M., Yu, F.R., Teng, Y., Leung, V.C.M., Song, M.: Computation offloading and content caching in wireless blockchain networks with mobile edge computing. IEEE Trans. Veh. Technol. **67**(11), 11008–11021 (2018)
10. Zhang, S., He, P., Suto, K., Yang, P., Zhao, L., Shen, X.: Cooperative edge caching in user-centric clustered mobile networks. IEEE Trans. Mob. Comput. **17**(8), 1791–1805 (2018)
11. Yu, D., Ning, L., Zou, Y., Yu, J., Cheng, X., Lau, F.C.M.: Distributed spanner construction with physical interference: constant stretch and linear sparseness. IEEE/ACM Trans. Networking **25**(4), 2138–2151 (2017)
12. Xu, J., Chen, L., Zhou, P.: Joint service caching and task offloading for mobile edge computing in dense networks. In: 2018 IEEE Conference on Computer Communications, INFOCOM 2018, 16–19 April 2018, pp. 207–215 (2018)
13. Wang, Z., Li, F., Wang, X., Li, T., Hong, T.: A wifi-direct based local communication system. In: 2018 IEEE/ACM 26th International Symposium on Quality of Service (IWQoS), June 2018, pp. 1–6 (2018)
14. Li, F., Wang, X., Wang, Z., Cao, J., Liu, X., et al.: A local communication system over Wi-Fi direct: implementation and performance evaluation. IEEE Internet Things J. (2020). https://doi.org/10.1109/JIOT.2020.2976114
15. Bai, B., Wang, L., Han, Z., Chen, W., Svensson, T.: Caching based socially-aware D2D communications in wireless content delivery networks: a hypergraph framework. IEEE Wirel. Commun. **23**(4), 74–81 (2016)
16. Wu, D., Liu, Q., Wang, H., Yang, Q., Wang, R.: Cache less for more: exploiting cooperative video caching and delivery in D2D communications. IEEE Trans. Multimedia **21**(7), 1788–1798 (2019)
17. Zhang, W., Wu, D., Yang, W., Cai, Y.: Caching on the move: a user interest-driven caching strategy for D2D content sharing. IEEE Trans. Veh. Technol. **68**(3), 2958–2971 (2019)
18. Paakkonen, J., Barreal, A., Hollanti, C., Tirkkonen, O.: Coded caching clusters with device-to-device communications. IEEE Trans. Mob. Comput. **18**(02), 264–275 (2019)
19. Asheralieva, A., Niyato, D.: Game theory and Lyapunov optimization for cloud-based content delivery networks with device-to-device and UAV–enabled caching. IEEE Trans. Veh. Technol. **68**, 10094–10110 (2019)
20. Li, J., et al.: On social-aware content caching for D2D-enabled cellular networks with matching theory. IEEE Internet Things J. **6**(1), 297–310 (2019)
21. Wang, J., Cheng, M., Yan, Q., Tang, X.: On the placement delivery array design for coded caching scheme in D2D networks. IEEE Trans. Commun., 12 (2017)
22. Jiang, L., Feng, G., Qin, S.: Cooperative content distribution for 5G systems based on distributed cloud service network. In: 2015 IEEE International Conference on Communication Workshop (ICCW), June 2015, pp. 1125–1130 (2015)
23. Kai, Y., Wang, J., Zhu, H.: Energy minimization for D2D-assisted mobile edge computing networks. In: ICC 2019–2019 IEEE International Conference on Communications (ICC), May 2019, pp. 1–6 (2019)

24. Zhang, X., Zhu, Q.: Distributed mobile devices caching over edge computing wireless networks. In: 2017 IEEE Conference on Computer Communications Workshops (INFOCOM WKSHPS), May 2017, pp. 127–132 (2017)
25. Hayat, O., Ngah, R., Hashim, S., Dahri, M., et al.: Device discovery in D2D communication: a survey. IEEE Access **7**, 131114–131134 (2019)
26. Blasco, P., Gündüz, D.: Learning-based optimization of cache content in a small cell base station. In: 2014 IEEE International Conference on Communications (ICC), June 2014, pp. 1897–1903 (2014)

Author Index

Printed in the United States
By Bookmasters